W0246874

Proficiency Testing
in Analytical Chemistry

Richard E. Lawn
Laboratory of the Government Chemist, Teddington, UK

Michael Thompson
Department of Chemistry, Birkbeck College, London, UK

Ronald F. Walker
Laboratory of the Government Chemist, Teddington, UK

THE ROYAL SOCIETY OF CHEMISTRY Information Services

LGC Laboratory of the Government Chemist

VAM VALID ANALYTICAL MEASUREMENT

A catalogue record of this book is available from the British Library

ISBN 0-85404-432-9

©LGC (Teddington) 1997

Published for the Laboratory of the Government Chemist
by The Royal Society of Chemistry,
Thomas Graham House, The Science Park, Cambridge CB4 4WF

Typeset by
Land & Unwin (Data Sciences) Ltd, Bugbrooke, Northants NN7 3PA
Printed by Athenaeum Press Ltd, Gateshead, Tyne & Wear, UK

Preface

This book has been produced as part of the Valid Analytical Measurement (VAM) Initiative, a programme funded by the UK Department of Trade and Industry. The VAM Initiative seeks to improve the quality of analytical data and to facilitate the mutual recognition of analytical results by promoting six key principles of good analytical practice. The VAM principles are:

1. Analytical measurements should be made to satisfy an agreed requirement.
2. Analytical measurements should be made using methods and equipment which have been tested to ensure that they are fit for their purpose.
3. Staff making analytical measurements should be both qualified and competent to undertake the task.
4. There should be a regular and independent assessment of the technical performance of a laboratory.
5. Analytical measurements made in one location should be consistent with those made elsewhere.
6. Organisations making analytical measurements should have well-defined quality control and quality assurance procedures.

The present work addresses the requirement for a regular and independent assessment of technical performance set out in principle 4, such assessments being one of the prime objectives of proficiency testing schemes.

The purpose of the book is to provide information on the proper use of proficiency testing as an analytical quality assurance measure, to all parties with an interest in the production and use of valid analytical data. As such it considers separately and specifically the needs and roles of the following groups:

- organisers of proficiency testing schemes;
- laboratories participating in proficiency testing schemes;
- end-users of analytical data.

Because of this approach there is some duplication in the coverage of certain aspects of the subject. For example, the role and objectives of proficiency testing are discussed in terms of organisers of schemes, participants in schemes and the ultimate end-users of analytical data, *i.e.* the customers of analytical laboratories. It is hoped that this feature will prove acceptable, enabling the reader with a specific interest in proficiency testing, whether as organiser, participant or end-user, to access readily those parts of the text of most relevance.

The material in this book has been enhanced by unique information obtained from extensive survey and investigative work carried out by the authors as part of the VAM Initiative. The surveys involved detailed discussions with the organisers of 18 UK proficiency testing schemes and an evaluation of questionnaire responses from over 200 participants in 13 UK schemes. The authors are very grateful to those organisations and individuals that contributed to the study.

It is hoped that the book will assist in the operation of proficiency testing schemes that actively promote the production of valid analytical data. In order to achieve these objectives, the essential principles of valid analytical measurement must be incorporated into the organisational structure of proficiency testing schemes. For example, the following aspects are emphasised: (i) the importance of establishing the traceability and uncertainty of the property values assigned to test materials used in proficiency testing schemes; (ii) the need for laboratory performance to be 'fit for purpose', which is not necessarily the same as stipulating that the highest possible accuracy must be obtained; and (iii) the role of proficiency testing in reinforcing other key components of VAM, such as the production of reference materials, the evaluation of analytical methods and internal quality control.

The design features of an effective proficiency testing scheme are described, but it is, of course, recognised that difficulties may sometimes arise in adopting certain features in practice. The options available for scheme operation are presented and their relative merits and shortcomings are discussed. The underlying philosophy is to assist in the operation of schemes to the highest standards possible within the constraints of any particular test sector. The types of scheme covered are those dealing with chemical analyses where a quantitative result is produced that is expressed either as a concentration (*e.g.* mg kg^{-1}) or as some other chemical measurement on a continuous scale (*e.g.* pH)

A further role of this book is to give particular attention to those aspects of proficiency testing schemes that are critical in securing improvements in the quality of analytical results. Effective feedback mechanisms between scheme organiser and participant laboratory are therefore discussed. It is also hoped that the information presented will assist in the implemetation of existing guides and protocols on proficiency testing, such as the International Harmonized Protocol for the Proficiency Testing of (Chemical) Analytical Laboratories and ISO Guide 43 (Proficiency Testing by Interlaboratory Comparisons[47]). The book is able to present a wider and more general discussion of the important principles, and the side issues, of proficiency testing than is possible in the necessarily limited confines of international standards.

The work is envisaged as having an educational role for analytical chemists, end-users of analytical data and students of analytical chemistry, in respect of the contribution that proficiency testing can make to the quality of analytical data. In particular, we seek to encourage analysts to participate in proficiency testing schemes and to do so in an appropriate and effective manner.

Finally, the book concludes with a brief 'forward look' at some of the issues likely to be of importance to proficiency testing in the future.

Richard E. Lawn, *Laboratory of the Government Chemist*
Michael Thompson, *Department of Chemistry, Birkbeck College*
Ronald F. Walker, *Laboratory of the Government Chemist*

August 1996

Proficiency Testing in Analytical Chemistry

Contents

Acknowledgements

The authors gratefully acknowledge the help of those proficiency testing schemes, listed in Appendix 2, that participated in the survey work. Special thanks are due to the Workplace Analysis Scheme for Proficiency (WASP), the Regular Interlaboratory Counting Exchanges (RICE), the Food Analysis Performance Assessment Scheme (FAPAS), the UK National External Quality Assessment Scheme (UK NEQAS) for clinical chemistry, the Brewing Analytes Proficiency Scheme (BAPS) and the CONTEST scheme, for permission to publish data and diagrams from these schemes. In the case of FAPAS, permission to reproduce Crown Copyright material was given by the Ministry of Agriculture, Fisheries and Food, CSL FAPAS® Secretariat.

CHAPTER 1

Proficiency Testing in the Context of Valid Analytical Measurement

1.1 What is Proficiency Testing?

A proficiency testing (PT) scheme comprises the regular distribution of test materials to participating laboratories for independent testing. The results are returned to the organiser of the scheme who makes an analysis of the results and reports them to all of the participants. The primary function of the scheme is to assist the participants to detect shortcomings in their execution of the test procedures and apply suitable remedial measures to make up any deficiency. This is a particularly important function in analytical measurement, which is fraught with practical difficulties and prone to unsuspected errors. As well as this primary 'self-help' ethos, there is also an element of accreditation in many proficiency testing schemes. Accreditation agencies will normally expect candidate laboratories to participate satisfactorily in a proficiency test in areas of analytical measurement where one is available.

Proficiency tests in analytical chemistry usually comply with a particular type, where the distributed materials are portions taken from an effectively homogeneous bulk material that resembles the normal test materials as closely as possible. The materials are distributed for analysis with choice of methods open, although participants should use their usual procedures. The estimated true value of the measurand is not revealed to the participants until after the collection of all of the results. The proficiency test therefore provides for the participant an independent check on the accuracy of the analytical result, albeit only at the time of the test. As the scheme is a regular event, it allows a laboratory to compare its results (i) with an external standard of quality, (ii) with the results of its peers and (iii) with its own past performance. Participation in a proficiency testing scheme therefore results in a much higher proportion of the participant laboratories reaching a satisfactory standard than would otherwise be the case. However, no consistent long-term improvements in performance will result unless the participation is in the context of an integrated quality assurance system within each laboratory.

Care must be taken not to confuse proficiency testing with other types of interlaboratory study, that is those designed for validating analytical methods or certifying reference materials. The operational protocols for these three types of study are quite distinct (Section 4.3).

1.2 Importance of Valid Analytical Measurements

Accurate and dependable analytical measurements are essential requirements for sound decision-making on many issues of vital interest to society. Environmental scientists and regulatory authorities need reliable data on the nature and extent of environmental contamination in order to identify cost-effective pollution-abatement technologies. The health and safety of the working population is safe-guarded by a number of regulations that prescribe maximum permissible levels of hazardous substances in the workplace environment. The effective enforcement of these regulations depends critically on the availability of sound analytical data. The welfare of the population at large is dependent on water and food supplies of high quality and this quality can be checked only by the careful application of a variety of different analytical techniques, each of which is often of considerable technical complexity, and therefore prone to error.

When valid measurements are not realised, data of poor quality are reported by a laboratory. In such circumstances the ensuing problems and costs for the end-user of the analytical data can be substantial and are associated with such consequences as:

- the costs involved in repeat measurements to correct poor data;
- the faulty decision-making that ensues when invalid results are acted upon;
- damage to reputation and credibility that results when an end-user is associated with poor data;
- possible loss of business where the end-user's customer is compromised by poor data;
- any legal and financial liability incurred from the use of poor data.

Hence it is important that those who commission laboratories to undertake analytical measurements on their behalf appreciate the critical need to select only competent laboratories.

There are a number of key concepts and practices that laboratories can apply to help them to improve, maintain and demonstrate the validity of their data, namely the use of properly validated methods, the use of internal quality control procedures, the use of certified reference materials, third party accreditation and participation in proficiency testing schemes.

1.3 The VAM Proficiency Testing Project

The material in this book has been enhanced by unique information gathered by the authors during a project on proficiency testing completed in 1994 as part of the UK Government's Department of Trade and Industry's Valid Analytical

Measurement (VAM) Initiative. This wide ranging project studied a number of important aspects of proficiency testing (PT). An efficacy study was undertaken to evaluate the effectiveness of proficiency testing as a means of improving the quality of analytical data and the essential features of effective schemes were identified. An investigation was carried out to establish the degree to which good performance in one particular test in a PT scheme correlates with good performance in a second, different, test within the same scheme. Such information contributes to the design of cost-effective schemes, by enabling optimum information on analytical proficiency to be obtained on the basis of a laboratory's performance in a limited number of critically selected tests. Other factors relating to the costs, benefits and resource requirements associated with proficiency testing were also reviewed.

The structure and operation of several UK proficiency testing schemes were surveyed, from the perspectives of scheme organisers, participant laboratories and end-users of analytical data (*i.e.* customers of analytical laboratories). Organisers of 18 schemes in the fields of chemistry and microbiology took part in the survey, which was conducted by means of face-to-face interviews and questionnaires; the participant survey secured questionnaire returns from over 200 participants in 13 different schemes. The proficiency testing schemes involved in these surveys are listed in Appendix 2. Telephone interviews were used to carry out a survey of 61 individuals classed as end-users of analytical data. The objective of the survey work was to identify current best practice and to pinpoint areas where improvements in the operation and use of proficiency testing procedures are required.

It was recognised at the outset that proficiency testing is only one of several quality assurance measures available to analytical laboratories. Consequently, the way proficiency testing interacts with and complements other quality assurance measures was investigated, so that the particular and unique role of PT could be identified and developed. A review of the statistical procedures required to handle the data generated in PT schemes was also undertaken, with the objective of identifying appropriate procedures for evaluating the performance of scheme participants.

In order to test or validate specific aspects of proficiency testing procedures recommended in this book, certain practical studies were carried out. Some of these involved collaboration with two proficiency testing schemes set up by the Laboratory of the Government Chemist as part of the VAM proficiency testing project. These schemes were the CONTEST scheme dealing with the analysis of contaminated land and the BAPS scheme dealing with the analysis of beer. The validation studies included a comparison of different procedures for establishing the assigned values of test materials; an assessment of the uncertainty associated with assigned values; an appraisal of the value of ranking laboratories according to their performance score, and the use of formal measures of efficacy to assess scheme effectiveness.

1.4 The Origin and Development of Proficiency Testing

Interlaboratory cooperation in pursuit of analytical quality began in earnest towards the end of the last century. Collaborative studies to validate analytical methods were being organised in the USA by the Association of Official Agricultural (now Analytical) Chemists in the 1880s and data from such studies have been reviewed by Horwitz.[1] Similar studies were underway in the UK by the early 1930s under the direction of the Society for Analytical Chemistry[2] and since 1980 this work has been continued by the Analytical Methods Committee of The Royal Society of Chemistry. In the early 1900s the first chemical reference materials (cast irons) certified for composition were produced in the USA by the National Bureau of Standards (now the National Institute for Science and Technology). In contrast to the above, interlaboratory programmes for the formal assessment of the proficiency of analytical laboratories are a rather more recent innovation.

In the field of clinical chemistry, where the analysis of human tissue has a direct impact on patient care, the first formalised survey of the proficiency of clinical laboratories took place in the USA[3] in 1947. The survey revealed large variations in results from different laboratories and, as a consequence, by the early 1950s the College of American Pathologists (CAP) had instituted regular proficiency testing in several laboratory disciplines in the health care area. A subsequent development in the USA has been the legislative requirement (*e.g.* Clinical Laboratory Improvement Amendments, 1988) that clinical laboratories participate satisfactorily in an appropriately designed proficiency scheme.

According to Bullock,[4] the first proficiency survey of UK clinical laboratories was reported in 1953 and revealed a wide spread of results for the determination of common constituents of blood. There followed other *ad hoc* surveys in the 1950s and 1960s, all of which confirmed the need for proficiency assessments on a regular basis. In 1969, the National Quality Control Scheme, in which test specimens prepared from human serum were distributed to the 200 participating laboratories every 14 days, was initiated by the Wolfson Research Laboratories, Birmingham, with funding from the UK's then Department of Health and Social Security.[5] Now known as the UK National External Quality Assessment Scheme (UK NEQAS) for general clinical chemistry, this scheme currently has about 600 participating laboratories. Concurrently, several other UK NEQAS activities have been developed, covering such areas as haematology, microbiology, immunology and drug assays. Within the last few years the funding of the UK NEQASs has shifted significantly to a self-financing basis, with operational costs now being covered by fees paid by participants. In addition to schemes such as the above which originated in the public sector, schemes have also been developed as commercial activities by private sector organisations. For example, in 1971 Wellcome Diagnostics (now Murex Diagnostics) initiated a clinical chemistry scheme, which was followed in 1978 by a scheme covering immunoassays.

The occupational hazard arising from exposure to asbestos dust has long been recognised and in order to assess the quality of asbestos monitoring a proficiency testing scheme was established in the USA in 1972, by the National Institute for

Occupational Safety and Health. A similar scheme was established in the UK in 1979 by the Institute of Occupational Medicine (IOM), following a recommendation of the UK's Advisory Committee on Asbestos.

Initially the activity was confined to a small group of specialist laboratories, but in 1984 the scheme, under the name of RICE (Regular Interlaboratory Counting Exchanges), was expanded to incorporate routine testing laboratories and it now has about 300 participants. A stimulus to participation in the RICE scheme is the legislative requirement of the Control of Asbestos at Work Regulations (1987) that laboratories undertaking asbestos fibre counting have the necessary facilities for producing reliable results.

The UK Control of Substances Hazardous to Health Regulations (1988) limit exposure to toxic chemicals in the work environment, and the proficiency testing scheme known as WASP (Workplace Analysis Scheme for Proficiency) deals with measurements of this type. WASP was set up in 1988 by the Health and Safety Executive and currently has about 200 member laboratories.

International legislation has also contributed to the development of proficiency testing in recent years. For example, the UK Ministry of Agriculture, Fisheries and Food initiated the Food Analysis Performance Assessment Scheme (FAPAS) in 1990, in part to meet UK obligations arising from various EU Directives covering the control of foodstuffs. A further impetus to the development of FAPAS was the evidence from *ad hoc* interlaboratory trials of poor quality analytical data for certain types of food analysis.[6] The AQUACHECK proficiency testing scheme for water analysis was established by the Water Research Centre in 1985 and the proficiency assessments reflect, in part, the requirements of relevant EU legislation on water quality.

Laboratory accreditation bodies have also encouraged the growth of proficiency testing schemes. In 1981, the National Association of Testing Authorities (NATA) in Australia set up a Proficiency Testing Advisory Committee and since then NATA have completed over 100 proficiency tests, some of which were one-off exercises, while others are on-going. In the UK, the UK Accreditation Service (UKAS) strongly endorses the role of proficiency testing in quality assurance and in some areas, for example asbestos testing, maintenance of a laboratory's accredited status is dependent on satisfactory performance in an appropriate PT scheme, such as RICE.

In addition to schemes such as those above that are run primarily on a national basis (although a small proportion of their participants may be from other countries), some schemes have been developed from their inception as international activities. Thus the scheme referred to as the Asbestos Fibre Regular Informal Counting Arrangement (AFRICA) and established in the mid-1980s by the IOM operates on a broadly similar basis to the RICE scheme, but has participants from over 30 different countries. The Wageningen Agricultural University in the Netherlands currently operates a number of international proficiency testing schemes concerned with the analysis of environmental samples such as soils, sediments and plant tissue. Participants number around 250 and are drawn from about 70 countries. With the increasing importance of the internationalisation of trade, environmental protection, law enforcement, consumer safety, *etc.* the need to demonstrate comparability of data between laboratories on an international basis will continue to grow and the need for fully international proficiency testing schemes can be expected to grow also. In this respect it is of interest to note that the EU, through its Standards, Measurements and Testing Programme, is currently (1995) proposing the establishment of a 'pan-European' proficiency testing network, aimed at supporting particular EU directives and regulations.

A further current development, stemming from the significant increase in proficiency testing activities in recent years, has been an international initiative to harmonise the operational procedures adopted by the organisers of proficiency testing schemes.[7,8] Agreement on the protocols to be followed is considered essential if all parties are to have the necessary confidence in the assessments of laboratory proficiency ultimately produced. In a further recent activity (June 1995) an expert group was set up by the International Organisation for Standardisation (ISO) to undertake an extensive revision of the ISO Guide on proficiency testing[9] first issued in 1984.

It is difficult to estimate the total number of proficiency testing schemes currently operating, but a UK listing[10] published in 1994 (see Appendix 1) recorded 26 different schemes, covering a wide range of measurement disciplines in the chemical and microbiological areas. For example, in addition to the schemes already referred to, there are schemes dealing with the analysis of such materials as fertilizers, animal feedstuffs, alcoholic beverages, contaminated soils, petroleum products, cereals and cane sugar. There are also schemes covering forensic investigations, the microbiological examination of foods and waters and veterinary investigations. It is possible that there are yet other schemes currently operating and the need to minimise duplication of effort in the establishment of new schemes means that there is a requirement for a comprehensive register of all existing proficiency testing activities. This particular issue is currently being addressed on a Europe-wide basis in the VAM 1994/1997 project on proficiency testing being carried out at the Laboratory of the Government Chemist.

The Role of Proficiency Testing in Analytical Quality Assurance

2.1 Objectives of Proficiency Testing

The detailed objectives of a proficiency testing scheme will depend on the scheme, the analytical sector and the test methods concerned. However, all PT schemes will share the following two key objectives:

(a) *The provision of a regular, objective and independent assessment of the accuracy of an analytical laboratory's results on routine test samples.*

(b) *The promotion of improvements in the quality (accuracy) of routine analytical data.*

The most important function of proficiency testing is to provide a regular and objective assessment of the quality of a laboratory's performance in its day-to-day work. To this end, test materials that resemble those routinely analysed are distributed on a regular basis (typically 3–12 times per year) to test laboratories: the assigned values for the test materials are not disclosed in advance to the laboratories taking part.

The second essential objective of proficiency testing is, where necessary, to help laboratories to improve the accuracy of their data. By means of proficiency testing, laboratories are able to discover any unsuspected inaccuracy they might have and, from their performance in subsequent rounds, review the effectiveness of the remedial actions they may have taken. Additionally, technical advice and guidance on analytical problems may be sought from scheme organisers. Where the overriding objective is to improve and maintain data quality, scheme organisers themselves may initiate approaches to laboratories experiencing persistent difficulties with a particular analysis.

In addition to the above, subsidiary objectives of proficiency testing schemes may include the following:

- provision of support to laboratory accreditation activities;
- the identification of competent laboratories for regulatory, commercial or other purposes;

- an assessment of the data quality of a particular test sector, rather than of individual laboratories;
- a comparison of the performance of different analytical methods used by participant laboratories for a particular measurement;
- the production of materials suitable for training purposes or quality control.

2.2 Aspects of Analytical Quality Assessed in Proficiency Testing

The basic concern of proficiency testing is accuracy – the closeness of agreement between a test result x submitted by a participant and the assigned value x_a set by the organiser. Clearly any inaccuracy contains a systematic aspect (bias), and a random aspect characterised by the precision of the measuring procedure. However, PT schemes do not normally address the magnitude of this precision separately. The participant laboratory can easily determine for itself whether imprecision or bias is the main determinant of its inaccuracy.

In practice each laboratory is scored on accuracy for each analyte in each test material in each round of the test. The error $x–x_a$ is converted into a score by the use of a scaling factor. PT schemes detect errors of a serious magnitude through the occurrence of abnormally high (or low) scores. The detection of such a (presumably) unexpected error acts as a warning to the participant that both the analytical procedure and the quality assurance measures undertaken in the laboratory are inadequate.

A PT scheme also addresses trueness in the sense of the ISO definition.[11] The average value of a large series of test results (*i.e.* those of all of the participants) can also be compared with the accepted reference value, although this is rarely an explicit part of the testing protocol. However, there is a more important sense of 'trueness' that impinges on proficiency testing. If the assigned value used is taken simply as the consensus of the participants' results in that round, then the scheme merely encourages the laboratories towards consistency. While such practice may in many circumstances produce a true result, in others it could institutionalise the use of inept analytical methods. Most analytical chemists embrace the ideal of striving towards the true value of the measurand (within the bounds set by uncertainty of measurement). This ideal can be realised in PT only by the use of an assigned value that is the best available estimate of the true value. Thus the choice of the procedure used to establish the assigned value is a critical feature in proficiency testing. This is exemplified by the experience of clinical proficiency testing in the USA, where participant consensus values for iron in serum were consistently low compared to the values found by a reference procedure.[12]

Other features of a laboratory's performance play a role in proficiency testing. For example, it is the ethos of PT that a participant laboratory should be using its routine methods of operation, so that the quality of its results reflects its normal output. Therefore the reported result should be in exactly the same form as would be supplied to a normal customer of the laboratory. There can therefore be no opportunity for the laboratory to correct inadvertent mistakes in reporting (such as transcription errors, use of wrong units, missing powers of ten in the result), any more than there could be in reporting results to a normal client.

In general, schemes set a strict deadline for the return of results by participants. This is essentially a matter of organisational convenience, however, and participants are not usually scored on timeliness, although some scheme organisers may wish to use this as one criterion for assessing competence.

2.3 The Applicability of Proficiency Testing

While, in principle, proficiency testing could be applied in most chemical measurement sectors, it is, by its nature, more readily applicable in particular circumstances. One important consideration, particularly for scheme organisers, is the minimum number of participating laboratories needed to ensure the technical viability of a proposed PT scheme. This factor will be especially important in those schemes using the mean of participants' results to determine the assigned values of test materials. When this is the case, the general view of organisers of proficiency testing schemes in the UK is that the minimum number of participants should be at the very least 10 and preferably 30 if a scheme is to be technically viable. However, where the assigned value is established independently of participants' results, it would be possible, in principle, to run a PT scheme for a single laboratory. Of course, an important feature affecting minimum size is finance. As most PT schemes are paid for from the fees paid by participants, another important consideration for organisers is the minimum participant numbers required to ensure the financial viability of a scheme.

From the viewpoint of participants, a proficiency testing scheme is most applicable as a quality assurance procedure where the laboratory has a large throughput of routine samples and tests of the type covered by the PT scheme. In these circumstances the laboratory can incorporate the PT materials into their routine work schedule without the need to make special arrangements for their analysis. This, of course, is how proficiency testing should always be tackled by analytical laboratories, so that the performance assessment obtained is a useful reflection of the quality of routine work. Where routine throughput is small, a laboratory may find that the costs of participation in a PT scheme are large in relation to the income they generate from their day-to-day work from the particular tests concerned.

Despite this, however, the survey of participants showed that there are instances where laboratories find proficiency testing applicable to their situation even though they have little routine work in the tests concerned. Such benefits as staff training, maintenance of expertise and demonstration of a commitment to quality were cited by these laboratories. Usually, therefore, proficiency testing is most readily applicable where a large number of laboratories (perhaps 30 or more) are regularly carrying out at least moderate numbers of routine analyses (perhaps 1000 or more annually). In such circumstances, the need for a regular assessment and demonstration of the quality of routine data is self-evident. Meeting such a need through proficiency testing is likely to be technically and economically feasible.

Proficiency testing is especially relevant to test sectors where reference materials are difficult or impossible to obtain. The materials distributed in a PT scheme, while of a lower metrological level than certified reference materials, are nevertheless well-characterised for the analytes of interest and therefore provide a vital substitute for CRMs where the latter are unavailable. In this context the establishment of schemes for environmental analysis, for example, might be considered of higher priority, in terms of promoting valid measurements, than the establishment of schemes for metallurgical analysis, where CRMs are readily available.

Proficiency testing is especially appropriate where a wide variety of analytical methods are in use for a given determination, since comparability of data would otherwise be difficult to demonstrate. However, where analytical methodology is non-routine or under development, proficiency testing is hardly applicable. In such cases there is no sound basis for establishing assigned values. Furthermore, a PT scheme could not be relied upon to give an accurate assessment of *laboratory* proficiency. Performance scores attributed to laboratories would be just as likely to reflect inherent problems in methodology rather than laboratory competence.

Finally, it may be noted that the nature of the test materials themselves may impose a limitation on the type of PT scheme that can be organised. Very hazardous materials (*e.g.* toxic, corrosive, explosive) might pose insuperable safety problems in relation to preparation and distribution of the materials. Likewise it may be impossible to organise proficiency tests for very labile analytes.

2.4 The Relationship of Proficiency Testing to Other Forms of Analytical Quality Assurance

The quality assurance of analytical data involves, amongst other activities, the use of several internal quality control (IQC) procedures by the analytical chemist. The overall aim of these procedures is to ensure that the analytical data routinely generated by laboratories are of acceptable precision and trueness.

Internal quality control addresses the problem of monitoring accuracy on a run-by-run basis, thereby ensuring that the data produced by any one laboratory are effectively bias-free and self-consistent. The establishment of a control chart for a particular analysis is an important and widely used mechanism that a laboratory can use to monitor the day-to-day consistency of its data output. At intervals during a run a portion of a material with known characteristics (the control material) is analysed and the results entered on the control chart. Any significant deviation of the observed result from the expected result may then be acted upon. To establish the repeatability precision of test data, duplicate analyses of a given proportion (perhaps 5–10%) of routine test materials may be undertaken (if all samples are not analysed in replicate as a matter of course). The spread in results between pairs of duplicate determinations enables the precision or repeatability of the data to be assessed.

With regard to the precision of data it should be noted that although a set of results may show close agreement within themselves, it does not automatically follow that the results are correct or true. A laboratory could be producing biased data consistently. While, in principle, internal quality control can provide a large measure of protection against the release of data that are not fit for purpose, IQC is not always carried out effectively. In such instances proficiency testing is especially important as it acts as an external check on data quality. A recent international initiative has established a protocol for the operation of internal quality control procedures.[13] Although internal quality control and proficiency testing are rightly regarded as independent procedures, the two are complementary and neither can operate fully effectively without the other. Proficiency testing is useful in demonstrating shortcomings not only in a participant's analytical methods, but in its IQC system as well. The main beneficial effect of proficiency testing on data quality in a laboratory probably results from its encouraging the participants to install effective IQC systems. This relationship has been demonstrated experimentally.[14] Although proficiency testing has this capacity for demonstrating shortcomings, it is IQC that acts as a check on the quality of data produced day-to-day.

To maintain the strictly external nature of proficiency testing as a QC measure, the assigned value of the test material and the expected interlaboratory variation of participants results should, wherever possible, be established by procedures that are independent of the participant data. It is usually preferable that assigned values are established by a group of specialist laboratories using well-validated analytical procedures, rather than relying on the consensus of participants results. It is quite possible (and indeed well known) for a participant consensus to be seriously biased through the general use of flawed methodology (*e.g.*, Figure 2.1). Another instance has been reported by Tietz *et al* [12] and concerns the determination of iron in human serum. Many routine methods fail to determine total iron due to incomplete recovery, so assigned values based on participant consensus will therefore be a poor estimate of the true iron-content. Consequently, laboratories using a more reliable method would receive a poor performance score, despite following better analytical practice. It is even more important that the acceptable

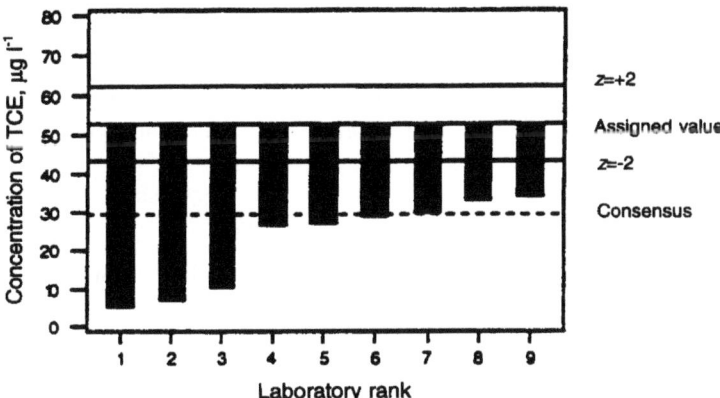

Figure 2.1 *Example of the consensus of participants being significantly different from the assigned value obtained by formulation – the determination of 1,1,1-trichloroethane (μg l^{-1}) in water. (Data from CONTEST)*

interlaboratory variation should be stipulated by the coordinator, in terms of fitness for purpose criteria, rather than calculated from participants' results.

It is important that analytical laboratories adopt a comprehensive QA programme that incorporates both internal and external QC features. In particular, laboratories should not rely exclusively on PT as a means of quality assurance as this will provide an incomplete picture of data quality. A comprehensive programme will include the following elements:

- the use of internal quality control;
- the use of validated analytical methods;
- the use of certified reference materials;
- participation in proficiency testing schemes;
- accreditation to an appropriate quality standard.

2.5 The Limitations of Proficiency Testing

Proficiency testing fulfils a necessary role in the production of data that are fit for purpose, but needs to be carried out in the context of the IQC practice outlined above. Like all such measures, proficiency testing has its own particular features and limitations that need to be considered when it is used as part of a laboratory's quality system.

(a) The retrospective nature of proficiency testing

The organisational structure of a PT scheme – the return of results to an external coordinator, the subsequent processing of results and the distribution of a performance report to the participant – entails an unavoidable time delay in the assessment of laboratory data quality. Proficiency testing is therefore retrospective in nature and performance scores reflect the quality of a laboratory's analyses at some fixed point in the past; several days to a few weeks may elapse before performance assessments are returned to participating laboratories. Under such circumstances it is not acceptable to rely exclusively on proficiency testing to ensure the day-to-day quality of a test laboratory's data. Internal quality control must also be employed. IQC then ensures laboratory consistency from run-to-run while the PT scheme provides a regular, but less frequent, external check of data accuracy.

(b) Characteristics of the proficiency test material

The suitability and quality of the test material distributed are important determinants of the effectiveness of a PT scheme. The proficiency test material should resemble the routine test materials closely and the assigned values for the various analytes must be reliably established, preferably by specialist laboratories using well-validated methodology. However, in some instances these requirements simply cannot be met, for a variety of reasons. In such circumstances the usefulness of the scheme may be compromised. For example, if it is necessary to prepare PT material by spiking a matrix with the analytes of interest (because a material containing native analytes at the appropriate concentrations is not available), caution will be required when drawing conclusions regarding laboratory performance on test materials. If the assigned values for the PT material have to be established from the consensus of participant results, it should be realised that such assigned values could be biased. Consequently, performance scores based on such a value could also be in error.

(c) The analyses covered by a proficiency testing scheme

The extent of proficiency testing is obviously limited by financial constraints. In laboratories that undertake a very wide range of analyses, only a small fraction of them can be subjected to a proficiency test. This fraction has to be regarded in some sense as representative of the general performance of the laboratory. There is no guarantee that that is a correct assumption.

The performance data provided by PT are often restricted to a small range of particular sample types (analyte/matrix combinations). However, it is not unknown for closely related tests to be conducted with different degrees of accuracy among a group of laboratories. Thus while a laboratory may perform well in the determination of zinc in a foodstuff by flame atomic absorption spectrophotometry, it does not automatically follow that it would perform equally well in the determination of calcium in the same foodstuff by the same technique. Figure 2.2 illustrates a well-documented example of this. On the other hand, within the organochlorine class of pesticides, the performance score obtained on one analyte often correlates well with the scores obtained on other pesticides. Hence, for performance on one analyte to be taken as indicative of likely performance on another analyte, a detailed and expert technical knowledge is required.

In the longer term, this specificity in the effect of proficiency tests may become less evident. Proficiency testing in specific tests should result in a general improvement of the IQC system in a laboratory, and that in turn should result in improvements in quality in all analytical procedures, even those not the subject of proficiency testing. However, there is no experimental evidence to support this conjecture at present.

(d) The disclosed/declared nature of proficiency testing

When laboratories participate in PT schemes, they are usually aware that they are being tested. In such circumstances there always exists the possibility that extra effort will be applied to the PT sample compared with that normally applied to routine samples. For example, a PT sample might be allocated to the most experienced analyst and/or the material might be analysed in duplicate where single analyses are the norm in routine work. Because of such factors, the reliability of proficiency testing as an indicator of routine performance depends upon the participating laboratories observing the spirit of proficiency testing, which is to treat PT samples in exactly the same manner as routine samples.

Fully blind or undisclosed trials can be arranged, but the extra effort and expenditure required of the scheme coordinator are considerable. Perhaps the most practical mechanism for ensuring that laboratories are accommodating PT samples

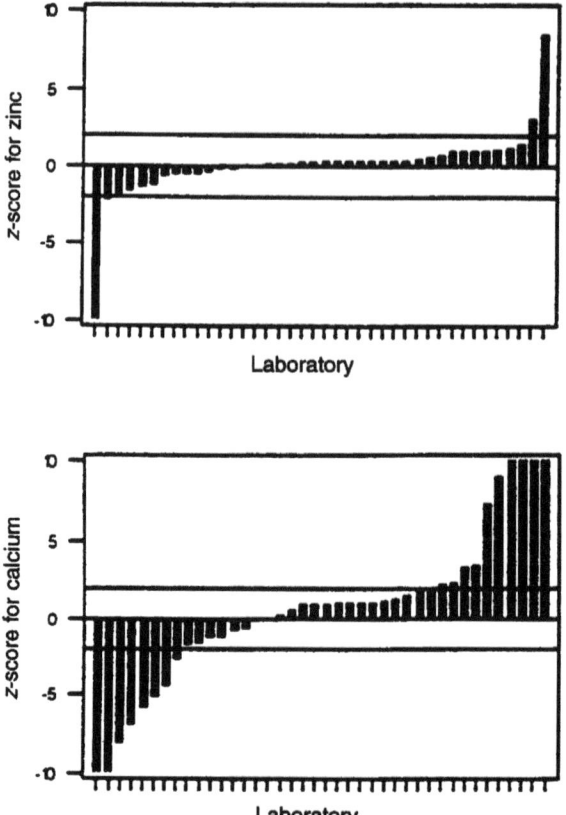

Figure 2.2   *z-Scores for zinc and calcium determined by the same participants on the same material, both by flame atomic absorption. This Figure shows that competence in determining an analyte does not necessarily imply competence in determining another by the same method. (Data from FAPAS)*

into their routine work schedule would be via quality auditing. Quality auditors could be encouraged to examine records with a view to establishing whether PT samples are being afforded special treatment.

CHAPTER 3

Performance Scoring in Proficiency Testing – An Overview

3.1 The Statistical Distribution of Participants' Results

In data collected under repeatability conditions there is a strong presumption, based on the Central Limit Theorem,[15] that the frequency distribution of the results will be close to a sample from a normal distribution. This is because we assume that there are numerous, small, independent errors made at the many stages of the manipulations in the analytical procedure. A combination of such errors tends to form the normal curve. In fact we usually find this presumption to be justified to a large degree where systematic effects (*e.g.*, drift) are absent.

However, PT data are not produced under repeatability conditions and often not even under reproducibility conditions, because several analytical methods or many variants of the same method may be used by the different participants in a round of the test. Consequently there is no strong *a priori* reason for the presumption of normality in PT data. Nevertheless, in the absence of disturbing influences, a bell-shaped distribution (like the normal) is often observed (Figure 3.1). However, it usually differs from the normal distribution in having 'heavy tails' (an unduly high proportion of observations far from the median) and in the presence of 'outliers' (results that are so far from the central tendency of the distribution to be more plausibly part of a different distribution). Therefore only the central part of such a distribution resembles the normal. However, the resemblance is usually sufficiently strong to suggest an interpretation of PT results based on the properties of the normal distribution, as most of the participants' scores will fall within this central range. The parameters of the central part of the distribution can be estimated by robust methods[16] (see also Section 5.14). These methods can be used to find the central tendency (robust mean) and dispersion (robust standard deviation) of the distribution without the influence of heavy tails and outliers, although the methods are applicable only to unimodal and roughly symmetric distributions.

15

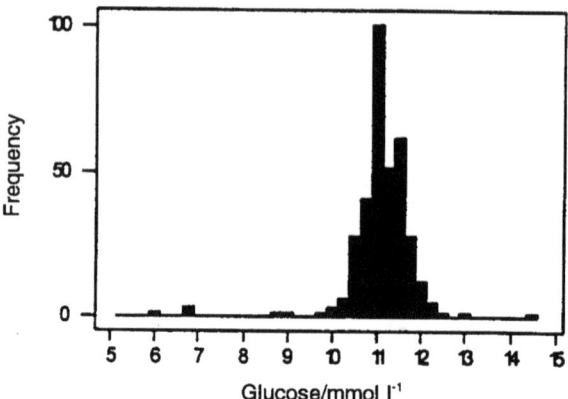

Figure 3.1 *Quasi-normal distribution of results (plus outliers) from a round of a proficiency test for the determination of glucose in blood serum. (Data from UK NEQAS)*

3.2 Approaches to Performance Scoring

Where x_a is the assigned value for the test material and x is the result reported by a participant, the size of the error $|x-x_a|$ is obviously dependent on the units in which the measurement is made. It will also vary with the difficulty of the determination and, to a degree, on the concentration of the analyte. Therefore some kind of scaling that would make the errors of equal magnitude in different situations would be helpful in allowing us to compare performances observed in different schemes and with different analytes. One obvious scaling of this type is to divide the error by the assigned value. This forms the basis for the 'q-score':

$$q = (x - x_a)/x_a$$

which is the relative deviation of the result. For a group of participants in a given round, q would tend to be zero-centred in the general absence of bias. However, the dispersion of q (*i.e.* its standard deviation) would tend to vary between analytes and matrices and, to a degree, for different concentrations of the same analyte (Figure 3.2).

A more natural scaling would be to divide the inaccuracy by a standard deviation σ to give the 'z-score':

$$z = (x - x_a)/\sigma$$

The parameter (σ) is known as the target value for standard deviation. A judicious choice of the value of σ would always result in a set of scores (from a number of participants in a round) that would be dimensionless, zero-centred and bounded approximately by a similar range (outliers aside). This could provide comparability between tests in a PT scheme and allow analytical chemists to comprehend immediately the meaning of a score from an unfamiliar PT scheme.

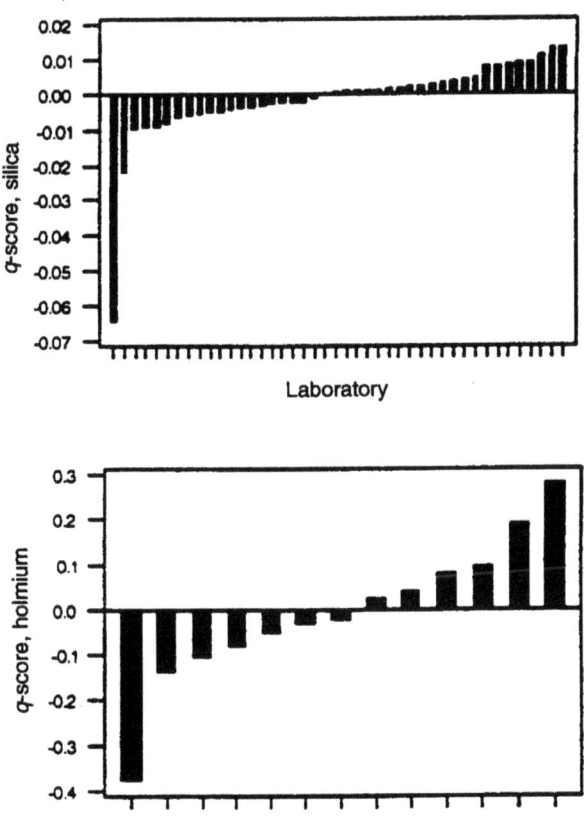

Figure 3.2 *q-Scores of participants in one round of a proficiency testing scheme, showing (top) a relatively small dispersion of scores for the analyte silica (SiO_2) and (bottom) a much higher dispersion of scores from the analyte holmium. This example illustrates that q-scores will not generally be comparable in magnitude among analytes. (Data from GeoPT)*

If the distribution of results from a round were actually normally distributed with a mean of x_a and a standard deviation σ then z would be the standard normal deviate. (This hypothetical situation is called 'well-behaved' in this book.) Very few z-scores would fall outside the bounds of ±3 (about 3 in 1000 on average) and about 5% would fall outside the bounds of ±2 in such a group. Hence in sets of z-scores from actual proficiency tests it would justifiable to regard such extreme z-scores as due not to random events but to the actual incidence of analytical problems, either bias or unacceptably large random error in the results. Therefore the z-score provides both a universal method of representing the results of a round and a method of providing decision limits for the proficiency testing scheme (Figure 3.3).

Some of the older established proficiency test schemes have more complex

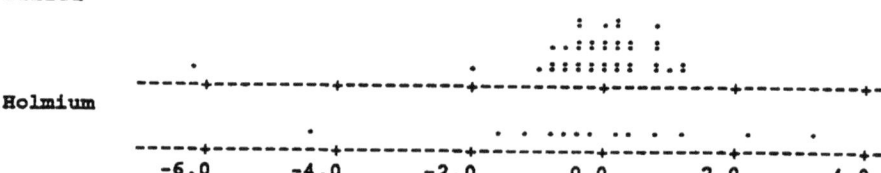

Figure 3.3 *Dotplots of z-scores for silica (%m/m) and holmium, calculated from the same results as the q-scores in Figure 3.2, showing that the scores have comparable values despite the great difference in concentration between the analytes (SiO₂ = 69.9%, Ho = 0.7 ppm). (Data from GeoPT)*

scoring systems. However, all of these are derived from either the z-score or the q-score, usually by arbitrary shifts and scalings.

3.3 Choice of Assigned Value

It is the normal purpose of analytical measurement to provide (within the bounds of measurement uncertainty) an estimate of the true result. The obvious choice of a value for x_a is therefore the best available estimate of the true value. This estimate can be obtained in a number of ways that are detailed elsewhere (Section 5.13). In some PT schemes it is recognised that the analytical method is empirical. Under such circumstances the consensus of the participants must necessarily be close to the true value and can be safely used as the assigned value. When the method is not empirical, the use of the participant consensus is not generally recommended because there is no guarantee that it is not biased. In some analytical methods there may be no harm in using the consensus. Examples of these methods arise with analytical determinations that are normally regarded as easy and where the principles and potential interferences of the method are well-understood and thoroughly documented. However, in difficult fields of analysis such as the determination of very low concentrations of pesticides and other toxins, it is not uncommon for the consensus to be biased (Figure 2.1), or even for there to be no real consensus among the participants (Figure 3.4).

In instances where the assigned value deviates from the consensus of the participants (given a normal distribution with standard deviation σ) there tends to be an excessive proportion of high absolute values of z-scores at one end of the distribution with a mode that does not coincide with zero.

3.4 Choice of Target Value of Standard Deviation

The choice of a value for σ, the 'target value for standard deviation', is straightforward in principle: the value should represent the maximum variability that is consistent with fitness for purpose. This criterion implies that the variability of the data (as represented by the PT results) would not impair the capability of an end-user of the data to make correct decisions based on the data. In practice,

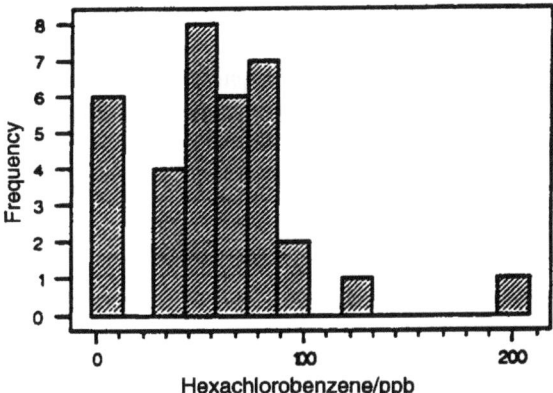

Figure 3.4   *Example of data from a round of a proficiency test (the determination of hexachlorobenzene in fat) where there is not a clear consensus among the participants. (Data from FAPAS)*

however, there is no general method for selecting a fitness for purpose value for σ. PT schemes in different sectors use criteria appropriate to that field, usually determined by the professional experience represented on the scheme's steering committee. Some schemes have used values of σ not based on fitness for purpose. Such values have been based on features such as (i) a general perception about the capabilities of the laboratories in a sector, or (ii) the dispersion of the actual results of the laboratories participating in a round. Such methods are merely descriptive of what is going on and are likely to be falsely reassuring to the analytical community. The fitness for purpose criterion therefore adds value to the resultant score that would otherwise be absent.

A general approach to selecting a value for σ based on minimal cost functions has been suggested[17] but not yet implemented in any specific sector of analysis.

When the chosen value of σ is smaller than the robust standard deviation of the results actually observed there will be an excessive number of z-scores with an absolute value of > 3 (given an unbiased assigned value): when σ is larger there may be few or even none. In practice a value of σ based on fitness for purpose rarely describes the observed dispersion at all exactly. This does not detract from the validity of classification or decision limits based on z-scores, because the value of σ is meant to be *prescriptive* rather than *descriptive* (see also Section 5.16).

3.5 Combination of Scores

There are a number of situations where the combination of several z-scores to produce an overall statistic seems attractive. For instance it might be useful to summarise the performance of a laboratory in a particular test over a period of a year. Often a participant will want to summarise the performance for several different tests in a round by means of a single statistic. There are various methods of combining z-scores that are statistically sound, and the resulting statistics have somewhat different properties. However, great care must be exercised in the use of these combined statistics to avoid incorrect conclusions or misleading statements. This admonition is of particular import when results are considered by those with no statistical training. Probably, a simple graphical display of the original z-scores would be more informative in such cases. However, with these reservations in mind, some approaches to computing combined scores are discussed in Sections 3.5.1 and 3.5.2 below.

3.5.1 The Rescaled Sum of z-Scores (RSZ)

This is defined as:

$$RSZ = \sum_i z_i / \sqrt{n}$$

where n is the number of scores to be combined. If the contributing z-scores can be regarded as standard normal deviates, then so can RSZ. Thus we could apply the same classification rules to RSZ as to individual z-scores. For example, only about 3 in 1000 values of RSZ derived from a well-behaved system would on average fall outside the bounds of ±3. A legitimate use of RSZ would be to summarise the performance of a participant in a specific test in a PT scheme in several rounds conducted over a period, say a year, for review purposes. Another such possibility is the combination of scores on the same analyte in different materials analysed in one round.

An RSZ value will tend to hide a small proportion of moderately high z-scores among mostly acceptable scores. For instance if we have the acceptable scores:

$$\{0.8, -0.9, -1.2, 1.8, -0.2, 1.4\}, \text{ then } RSZ_1 = 0.69$$

If we have similar sets with unacceptable z-scores embedded we might have:

$$\{0.8, -0.9, -1.2, 1.8, -3.9, 1.4\}, RSZ_2 = -0.82 \text{ or}$$
$$\{0.8, 5.0, -1.2, 1.8, -3.9, 1.4\}, RSZ_3 = 1.59$$

Both of the latter RSZ values seem acceptable but mask the presence of high absolute z-scores that indicate occasionally unsatisfactory performance. However, this is not necessarily a negative feature of RSZ if it is agreed that laboratories do not have to achieve satisfactory results every time in order to be considered competent. A more questionable use of RSZ is the combination of z-scores for different tests in the same round to provide a summary statistic. Such a practice could serve to disguise that one specific test out of several was *always* poorly executed. This can be seen in Table 3.1.

Table 3.1 *z-Score data for a single participant demonstrating that a combination score such as RSZ can often hide the fact that the participant is consistently not performing satisfactorily for a particular analyte (analyte 5)*

Round	Analyte 1	Analyte 2	Analyte 3	Analyte 4	Analyte 5	RSZ
1	0.43	−0.29	−1.15	1.28	1.52	0.8
2	−1.33	−0.54	1.9	1.33	3.43	2.14
3	0.13	−1	0.83	0.52	3.5	1.04
4	−0.22	0.51	−0.33	−1.39	2.96	0.68
5	0.55	−1.32	0.42	−1.4	1.7	−0.02
6	−1.08	1.2	−0.22	1.22	3	1.84
7	0.58	0.53	−0.23	0.25	2.99	1.84
8	0.63	−0.25	−0.27	−1.89	2.86	0.48
9	−0.47	0.56	0.84	1.47	−2	0.18
10	−1.07	1.38	0.06	−0.03	2.75	1.38

3.5.2 The Sum of Squared z-Scores (SSZ)

One of the problems noticeable in one example above is that high scores of opposite sign tend to cancel and allow the *RSZ* to remain at a small absolute value. A score that overcomes that defect is the sum of i squared z-scores given by:

$$SSZ = \sum_i z_i^2$$

which for a well-behaved system would have the chi-squared distribution with i degrees of freedom. Such a method of combining scores provides a compromise between bias and precision but, because of the square term, avoids the cancellation of large scores of opposite sign. For the example data sets in the previous illustration we have:

$$SSZ_1 = 8.1$$
$$SSZ_2 = 23.2$$
$$SSZ_3 = 47.5$$

The percentage points for the chi-squared distribution with six degrees of freedom are:

$$\text{for } p = 0.95, \ 12.6; \text{ and for } p = 0.995, \ 18.5$$

This clearly shows the improbability of SSZ_2 and SSZ_3 occurring in a well-behaved system.

A disadvantage of the *SSZ* is that it is rather sensitive to single outliers to allow for easy interpretation. A further problem is that there is no simple way of making the statistic commensurate for different degrees of freedom. In other words, it is difficult to compare an *SSZ* based on (say) three z-scores with one based on the combination of four z-scores. This could be an important drawback if it was required to compare the combined scores of a group of participants that had conducted different numbers of tests. Methods of transforming the chi-squared distribution to a normal curve with a good degree of approximation do exist and could be used to make different chi-squared distributions comparable, but they are beyond the scope of this book.

A seemingly attractive alternative combination score is the sum of the absolute z-scores. This sum would avoid cancellation of large z-scores of opposite sign and also undue sensitivity to single outliers. However, this statistic is mathematically intractable and is not recommended at present.

3.6 Classification of Participants

z-Scores are best used to alert participants to an unexpected source of error in the analytical system. Thus a value of $|z| \geq 2$ acts as a warning of potential problems, and a value of $|z| \geq 3$ is an indicator that remedial action must be undertaken. These criteria are best regarded as decision limits to be used by the participant to trigger remedial action. This is the principal purpose of proficiency tests.

It is also possible to classify z-scores on a similar basis. For example, a classification as follows could be proposed:

$$|z| \leq 2, \quad \text{satisfactory;}$$
$$2 < |z| < 3, \quad \text{questionable;}$$
$$|z| \geq 3, \quad \text{unsatisfactory.}$$

From a strictly technical viewpoint, such a process has little to commend it, as the original z-score provides more information than the resultant classification. A preferable approach is to regard a particular level of z-score as a trigger value, above which a feedback mechanism within the participant laboratory is activated to investigate and remedy the situation leading to the unacceptable z-score. Classification can seem to be beneficial in situations where those with no statistical training need to review and evaluate laboratory performance. However, it must be recognised that such situations are in themselves highly undesirable, from the standpoints of both the individual conducting the evaluation and the laboratory being evaluated. In particular, caution should be exercised when evaluating z-scores from a single round of a scheme (see also Sections 2.5, 5.18).

Classification of laboratories for licensing, accreditation or commercial purposes is sometimes based on the results of proficiency tests. The limitations of using combination scores for this purpose are clear from the discussion in Section 3.5. A preferred method is for the accreditation or licensing body to set an arbitrary minimum success rate based on several rounds of the test. For example, it might be appropriate to specify that a laboratory has to achieve a $|z| < 3$ in four out of the last five rounds for each analyte of interest. However, the range of circumstances that might be applicable here are so wide that this subject cannot be treated in a general way. Accreditation or licensing agencies will have to take into account the specific requirements of their field of activity in order to evolve a set of appropriate rules. These would have to take into account the relative importance of particular tests within each round, the likely consequences of incorrect data being released by a laboratory, and mechanisms for the re-accreditation of discredited laboratories.

3.7 Ranking of Participants

There is a strong tendency among participants (and to a lesser degree among PT organisers), at the end of each round, to want a list of participants that is ranked on the basis of the z-scores produced within the round. Such a list is widely perceived to represent the relative abilities of the participants, and to encourage low ranking participants to better performance. It is sometimes thought that in a commercial context such information would have potential as advertising or promotional value. There are two main reasons why ranking laboratories in this way is likely to be seriously misleading. Firstly, ranking in a round is necessarily based on a single combination score over several analytes, and such a score as seen above (Section 3.5) could mask a long term problem in some of the tests. Secondly, in typical PT schemes the ranking of a laboratory is likely to be very variable simply on statistical grounds. In other words, a laboratory could have a high, middling or low rank purely by statistical chance, whilst still exhibiting satisfactory performance. The rank achieved does not represent the competence of the laboratory, and is essentially incapable of being interpreted. This has been demonstrated in a computer simulation of a proficiency test carried out as part of the validation phase of the VAM proficiency testing project. The use of ranking is therefore not recommended in proficiency tests.

CHAPTER 4

Organisation of Proficiency Testing Schemes – General and Management Aspects

4.1 Competence of Scheme Organisers

Any body setting out to establish and operate a proficiency testing scheme must be able to demonstrate its competence if the scheme is to make a credible contribution to promoting data quality. This is especially important now that commercial operators are joining government organisations as organisers of proficiency testing schemes. The main requirements for organisers are shown in Box 4.1.

It is recognised that the necessary competence will require input to the scheme from several individuals, perhaps from different organisations. Accordingly, as discussed in Section 4.10, the effective operation of a scheme will require members of the steering committee or advisory committee to provide between them the complete set of skills required.

While the above itemises the main skills required in a PT scheme organiser, the operational procedures adopted by a scheme should be formally documented in both an operational protocol and a quality manual. The procedures applied in a scheme could then, if appropriate, be audited by an independent certifying body, so that the scheme's operation is registered to a quality system standard such as the ISO 9000 series.[18]

The VAM survey of UK PT scheme organisers (Appendix 2) has identified a range of benefits accruing to the organisers of schemes. These include:

- improved awareness of measurement problems;
- allows organisers to retain specialist staff/specialised equipment which would not otherwise be available;
- regular income stream from fees;
- income from sale of surplus test material;
- income from possible 'spin-off' consultancy work;
- allows development of databases of laboratories active in a given measurement sector;
- publicity.

Appendix 2: Proficiency Testing Schemes Taking Part in the VAM Survey

Agricultural Development Advisory Service (ADAS) Analytical Chemistry Monitoring Scheme.

AQUACHECK.*

British Petroleum InterCentre Precision Monitoring Scheme.*

Fertiliser Manufacturers Association Check Analysis Scheme.

Flour Milling and Baking Research Association.*

Food Analysis Performance Assessment Scheme (FAPAS).*

Forensic QA.

Murex Quality Assessment Programme.*

National Agricultural Check Sample Scheme (NACS).*

Proficiency Testing for Alcoholic Strength (ProTAS).*

Public Health Laboratory Service External Quality Assessment Scheme (PHLS EQAS) for Water Microbiology.*

Regular Interlaboratory Counting Exchanges (RICE).*

Sugar Association of London.

UK National External Quality Assessment Scheme (NEQAS) for general clinical chemistry.*

UK National External Quality Assessment Scheme for peptide hormones and tumour markers.*

Veterinary Laboratory Quality Assessment Scheme.*

Workplace Analysis Scheme for Proficiency (WASP).*

Further information on the schemes may be found in Appendix 1.

* Laboratories participating in these schemes also took part in the survey.

Box 4.1 *Main requirements for organisers of proficiency testing schemes*

- an expert knowledge of the analytical tests involved in the scheme and the analytical sector targeted by the scheme;
- the ability to prepare, on a regular basis, adequate quantities of good quality test material;
- the statistical expertise needed to evaluate test materials for homogeneity and stability, to evaluate participants' results and calculate performance scores, to evaluate the success of the scheme and to deal with contingencies;
- the ability to prepare clear performance reports, incorporating graphical presentation of data, for prompt distribution to participants after each round of the scheme;
- the resources to provide follow-up advice and guidance to participants who perform persistently poorly in the scheme;
- the ability to review the effectiveness of the scheme by, for example, monitoring trends in participant performance scores over a number of rounds; evaluating the validity of test material assigned values and evaluating the appropriateness of the criteria adopted for defining acceptable performance.

4.2 Scheme Objectives

The specific objectives of a particular proficiency testing scheme must be agreed and fully documented by the scheme organiser. All parties will then have a clear understanding as to what is being attempted and which protocols are being followed. The objectives of all schemes will include the following.

(a) The provision of a regular, objective and independent assessment of the quality of an analytical laboratory's results under normal conditions

Consideration will need to be given to the aspects of quality that need to be addressed. For quantitative analysis the concepts that define data quality are accuracy, trueness and precision. However, the basic concern of proficiency testing is an assessment of accuracy and consequently all schemes will compare a laboratory's reported result to the assigned value for the distributed test material concerned. This objective will be attained only where the assigned value of the test material is known to be a reliable (non-biased) estimate of the true value. Where a scheme relies entirely on the consensus of participants' results to establish an assigned value, without supplementary evidence to confirm the validity of the assigned value, the scheme may only be addressing conformity between the laboratories.

A further aspect to be considered by organisers is that 'quality' must be taken to mean that analytical data are *fit for purpose*. It does not necessarily follow that data have to be of the highest possible accuracy to be of acceptable quality. Organisers should therefore bear this in mind when setting the target standard deviation (Section 5.16) for a particular measurement, so that the objective of assessing the quality of routine data quality is achieved in a realistic, cost-effective manner.

In certain circumstances timeliness of reporting of results by participants may be a criterion by which quality is judged. Thus a laboratory failing to submit a result by a specified deadline would be deemed to have an unsatisfactory performance.

(b) The promotion of improvements in the quality (accuracy) of analytical data

Experience has shown that proficiency testing can be very effective in promoting improvements in accuracy (Section 4.12). However, certain operational features of schemes are critical if schemes are to fulfil this objective successfully.

Reports after each round of the scheme must be distributed promptly to participants and contain a clear presentation of the performance assessments. Such assessments should be transparent and provide a formal performance score that is based on simple statistics and preferably avoids the use of arbitrary scaling factors. However, for a scheme to be fully effective in promoting improvements in data quality, further feedback to participants, particularly those experiencing persistent difficulties with a particular determination, is required. Such feedback should be directed towards providing constructive technical advice and guidance to participants and could include the establishment of a formal advisory panel of analytical experts, regular open meetings for participants and workshops on analytical techniques.

Scheme organisers should also consider collecting information on the analytical methods used by participants and provide a supplementary report summarising performance scores in terms of the methods applied. Such an approach sometimes enables less reliable methodology to be identified and is therefore a further way by which scheme organisers can provide helpful information to laboratories producing poor results.

Participant laboratories should be encouraged to employ a comprehensive range of quality assurance measures, with emphasis on the importance of good internal quality control as a means of ensuring day-to-day consistency of data, the use of validated analytical methods, the use of certified reference materials to establish traceability of results to reliable measurement standards and audit of the quality system adopted by an independent accreditation body.

Any surplus test material remaining after the completion of a round is a valuable resource for internal quality control. It could usefully be offered to laboratories experiencing analytical difficulties for use as a training material or as a quality control material.

Subsidiary objectives of proficiency testing schemes include the following:

(c) The identification of competent laboratories for regulatory, commercial or other purposes

The formal identification of laboratories competent to carry out enforcement analysis in connection with legislation provides a further reason for operating a proficiency testing scheme. In that situation, an essential output of the scheme would be a published list of laboratories designated as competent. It is important that the way PT performance scores are used to designate competent laboratories is well-founded and clearly documented. Thus designation should not be based on a single performance assessment, but on performance over several rounds. It should also be appreciated that information on performance obtained from PT schemes is sometimes specific to particular tests (Section 2.5c).

(d) Provision of support to laboratory accreditation activities

Increasingly, participation in appropriate PT schemes will be a requirement for the accreditation or certification of test laboratories to quality system standards such as ISO Guide 25[19] and the ISO 9000 series.[18] The performance of laboratories in schemes and their response to any measurement problems revealed will be used as one criterion for awarding accredited status. Organisers of schemes and the appropriate accreditation agency should, therefore, agree and document the precise way in which PT performance will be reviewed and used for this purpose. ISO Guide 43, which is currently undergoing revision (due for completion in 1997), will provide comprehensive guidance on the role of proficiency testing in laboratory accreditation (see also Section 7.7).

(e) A comparison of the performance of different analytical methods

Where the scheme has a large number of participants using a variety of different methods for the same test, it may be worthwhile to report/collate performance scores in terms of the analytical methods applied. This provides an inexpensive means of comparing the accuracies of different procedures and enables less reliable methodologies to be identified and discarded. Schemes having this as an objective need to devise an appropriate means of collecting, collating and analysing the information, and reporting the outcome back to participants. Although this idea is attractive in principle, in practice the information collected on analytical procedures is often difficult to interpret. The idea works best when complete standard methods are being compared, rather than variations among a family of related procedures. An example of the benefits of this approach is provided by UK NEQAS and relates to the determination of lead in blood.[20] Compared with other methods, the dithizone procedure was found to be consistently less reliable.

4.3 Proficiency Tests and Other Interlaboratory Studies

Proficiency testing is just one type of interlaboratory activity. Two other types are: collaborative trials (or method performance studies) organised to validate analytical methods, and certification trials organised to characterise reference materials. These three activities are quite different in their objectives and operational protocols (Table 4.1). In a collaborative trial it is a proposed analytical method that is being studied and all participating laboratories must use this method exactly as specified by the trial organiser. Such a condition does not apply

Table 4.1 *Proficiency testing compared to other interlaboratory exercises*

Interlaboratory exercise	Protocols/guidelines	Objectives	Participants	Method of analysis used	Frequency of exercise
Proficiency testing	References 7, 47	To assess the competence of analytical laboratories	Open to any laboratory	Participants use a method of their own choice	At regular intervals; typically every 2 weeks to every 4 months
Collaborative trial	References 40, 41	To establish the precision and/or trueness of an analytical method	Experienced laboratories only	Specified exactly by the organiser of the trial	One-off exercise
Certification exercise	Reference 29	To characterise and certify a reference material	Experienced laboratories only	As agreed between organiser and participants. At least two different methods preferred	One-off exercise

in proficiency testing, where the performance of a laboratory in the application of its routine test methods is being assessed; such routine methods may well vary from laboratory to laboratory. Also, as a function of a method validation is to establish the repeatability of the method, the number of replicate measurements to be carried out will be specified. Again this stipulation does not apply in proficiency testing, as the participant laboratory would be expected to use its routine procedures, which may require single or duplicate analysis, depending on the application or laboratory concerned.

In a reference material certification study the objective is to form an estimate of the true value that is as accurate as possible. Consequently, participation is restricted to specialist laboratories using well-researched, validated and preferably definitive test methods. Again such restrictions do not apply to the participants in a PT scheme.

For reasons of the above type, therefore, the organisation of an interlaboratory trial for the purpose of proficiency testing cannot be combined with a method validation or formal reference material study. It is important that organisers and participants in a proficiency testing exercise understand this and appreciate the true objectives and methods of proficiency testing. IUPAC have published some useful guidance on the essential features of different types of interlaboratory study and their nomenclature.[21]

4.4 Resources Required to Operate a Scheme

The tasks involved in the operation of a proficiency testing scheme may be considered to fall within four main areas of activity. The tasks within each area are listed in Box 4.2.

The resources required to carry out these tasks can be summarised under the following headings:

- (*a*)   staffing requirements;
- (*b*)   consumables;
- (*c*)   specialised equipment;
- (*d*)   accommodation;
- (*e*)   specialist expertise.

(*a*) *Staffing Requirements*

The total effort required to operate a scheme varies greatly from scheme to scheme as it is dependent on a complex combination of factors, *e.g.*, total number of tests covered, number of tests per sample, number of participants (and whether they are national or overseas), frequency of distribution, complexity of sample preparation, *etc*. From the VAM survey of 18 UK scheme organisers (Appendix 2) it was found that the total effort ranges from less than 0.01 to 7 person years per year, a typical figure being 1 person year per year which could be expected for a scheme with 75 participants, covering 15 different tests and distributing samples on a monthly basis.

Box 4.2 *Main activities required to organise a proficiency testing scheme*

(i) *Sample preparation and evaluation*:
  - purchase of material;
  - preparation of material;
  - homogeneity/stability testing *etc.*;
  - establishing the assigned value of the analytes under test.

(ii) *Data handling*:
  - statistical analysis of all results;
  - calculation of performance scores;
  - assessment of performance scores;
  - preparation of documentation and reports.

(iii) *Liaison with participants:*
  - distribution of samples and instructions;
  - distribution of reports;
  - dealing with enquiries;
  - advising on analytical problems.

(iv) *Management*:
  - publicity;
  - internal documentation and review;
  - co-ordination of the scheme;
  - archiving of data.

The breakdown of the total staff effort required to operate a scheme also varies greatly from scheme to scheme and is again dependent on a wide variety of factors, *e.g.* whether the preparation and/or evaluation of the distributed materials is carried out 'in-house' or by a sub-contractor, complexity of sample preparation, number of participants and whether it is the organiser's policy to liaise with participants directly regarding analytical problems *etc.* According to the survey, however, typical allocations of effort in the four work areas might be:

Material preparation	~35%
Data handling	~35%
Liaison with participants	~15%
Management	~15%

In addition to the above staffing levels involved in the day-to-day operation of a scheme, there will be a small staffing requirement for servicing the scheme's steering board. The survey of UK schemes showed that, typically, steering boards have between four and sixteen members and meetings occupy about three days per year.

(b) Consumables

A wide range of consumables is likely to be required for the organisation of a particular scheme. These can be broken down as shown in Box 4.3.

Box 4.3 *Consumables needed in organising a proficiency testing scheme*

  (i) *Sample preparation and evaluation*:
- reagents;
- test materials/analytes;
- storage containers/labels;
- general laboratory consumables.

 (ii) *Data handling*:
- computer disks;
- printer paper;
- copier paper.

(iii) *Liaison with participants*:
- packaging;
- labelling;
- postage;
- paper;
- disks;
- telephone.

(iv) *Management*:
- paper;
- disks for archiving data.

 (v) General:
- photocopying;
- heating;
- lighting.

Data collected during the VAM survey of scheme organisers showed that expenditure on consumables might typically be £15k per annum.

(c) Specialised Equipment

Specialised equipment and facilities will be required to operate a PT scheme. In general, all schemes need similar facilities for the treatment of results, liaising with participants and management, namely:

- packaging facilities/arrangements if specialist packing is required for distribution of samples;

- labelling facilities;
- franking machine;
- computers and software;
- printer;
- photocopier;
- telephone/fax machine.

However, differences in requirements will occur, mainly in the area of sample preparation and evaluation. Each organiser requires laboratory facilities specific to the particular area of analytical chemistry with which they are involved, as well as general laboratory facilities. If all preparation and evaluation of the distributed materials is carried out by sub-contractors, then laboratory facilities need not be a consideration of scheme organisers.

Examples of the specialised equipment for test material preparation and homogenisation reported by scheme organisers include apparatus for:

- grinding;
- homogenisation;
- lyophilisation (freeze drying);
- sample division;
- sample dispensing;
- incubation;
- heat sealing of packaged materials.

(d) Accommodation

The types of accommodation required to organise a PT scheme may include:

- laboratories;
- offices;
- computer rooms;
- storage (for samples, including cold storage where necessary);
- storage (for archiving records *etc.*).

However, these areas may not necessarily be dedicated to the organisation of the PT scheme alone. The accommodation may already be in place for other activities of the organiser, *e.g.* the provider of an analytical service or the producer of chemicals. The VAM survey showed that although six of the eighteen scheme organisers surveyed have no accommodation dedicated solely to the organisation of their PT schemes, the other schemes do have dedicated space. This ranges from between one and four offices, and one and three laboratories, as well as specialised computer rooms and storage areas.

(e) Specialist Expertise

For many aspects of scheme organisation, staff need to be trained and experienced in the particular analytical field to which the scheme relates. This is particularly important in the areas of preparation and evaluation of the proficiency test materials and in liaison with participants regarding analytical problems and other queries. Liaison with participants also requires telephone and good communication skills.

In the area of data handling and production of reports, staff require keyboard and computing skills and, in particular, confidence in the use of statistics. The degree of experience will clearly depend on the extent to which complex statistics are used in the assessment of the results. In some cases, specialist statistical advice may be required to develop appropriate software when a scheme is first established. A statistical expert must always be available as a member of the steering committee, as unforeseen outcomes are not uncommon.

A range of skills are required by staff involved in the management aspects of a PT scheme, including organisational skills, knowledge of statistics, computer skills, accounting/book-keeping skills and the ability to work under pressure and to deadlines.

4.5 Essential Features of an Effective Scheme

PT schemes that are effective in promoting improvements in participant performance will possess the following features and organisers should pay special attention to them.

(a) *Management of the scheme*

1. The scheme should have a written protocol describing its operational procedures.
2. The objectives of the scheme should be clearly stated and the success of the scheme in meeting these objectives should be regularly reviewed.
3. Advice on the organisation and operation of the scheme should be available from a steering committee or similar body, whose members include representatives experienced in the tests concerned.

(b) *Distributed material*

4. The materials should be of good quality (*e.g.* sufficient homogeneity and stability).
5. The materials should be sufficiently similar to the routine test materials analysed by participant laboratories (*e.g.* matrix composition, physical form, analyte speciation, analyte concentration).
6. The procedures used to establish the assigned value of proficiency test materials must be documented and available to participants.

(c) *Assessing and reporting participants' performance*

7. The procedure used to define what constitutes acceptable performance should be documented.

8. The system used for scoring performance should be clearly explained and documented. The target value for the standard deviation should be available to participants before the proficiency test is conducted.

9. Reports on performance should be returned to participants promptly after each round of the scheme.

(d) Liaison with participants

10. Clear instructions must be given to participants regarding the analysis of the test material and the reporting of the results to the scheme organiser.

11. There should be a mechanism for establishing participant opinion on how well the scheme is meeting their needs.

4.6 Additional Features of a Scheme Based on VAM Principles

The VAM Initiative emphasises certain key aspects of data quality that must be addressed by all analysts if meaningful results are to be produced. For example, analytical methods and equipment should be tested or validated to ensure that the results they produce are fit for purpose; certified reference materials or calibration standards, traceable to a recognised authority, should be analysed to ensure valid data and to promote comparability between laboratories. Measurements should be made to a specification agreed with the customer, which should include the acceptable uncertainty in analytical data. Quality procedures should be subject to independent approval by an accreditation or certifying body.

Apart from analysts complying with these principles, it is equally important that proficiency testing schemes are also organised in accord with them. A scheme that reflects the essential principles of valid analytical measurement will possess the following features, as well as those listed in Section 4.4:

(a) Assigned values

1. Assigned values of test materials should wherever feasible be established by validated/definitive procedures carried out by specialist laboratories.

2. Assigned values should if possible be traceable to international measurement standards.

3. The uncertainty attaching to assigned values should be established.

(b) Review of participant performance

4. The criteria used for defining what constitutes acceptable performance (data accuracy) should be set in terms of fitness for purpose, *i.e.* end-user requirements.

5. z-Scores should be used to quantify participant performance.

6. Organisers should periodically review the trends in participant performance observed over several rounds of the scheme.

   7. Organisers should consider appropriate follow-up action in respect of persistently poor performers.

(c) *Review of method performance*

   8. z-Scores should be reviewed in terms of the particular analytical methods used by participants, in order to help characterise the performance of the various methods.

(d) *Test materials*

   9. Organisers should seek to develop surplus test material as reference materials or quality control materials.

(e) *Liaison with participants*

   10. Organisers should provide an advisory service for laboratories experiencing analytical difficulties with particular tests.
   11. Organisers should encourage participating laboratories to operate a comprehensive quality assurance programme.

(f) *Review and publication of scheme performance*

   12. Schemes should have a quality manual and organisers should seek certification of their schemes to an external quality system standard.
   13. Whilst preserving participant confidentiality, there should be regular publication of general performance information and trends in the open scientific press in order to bring the lessons learnt to the widest possible audience.

4.7 Participant Numbers

For a scheme to be both technically and financially viable there will usually be certain minimum participant numbers that scheme organisers should aim for. It is impossible to give precise figures, but the following guidance is offered, based on the opinion of organisers of currently operating UK schemes.

In respect of technical viability, the issue of participant numbers is most important where participants' results on test materials are used to establish assigned values. Ten participants is an absolute minimum if a reasonable estimate is to be made by this method.[†] If a scheme adopts a different procedure for arriving at assigned values (*e.g.* by use of one expert laboratory, by using a certified reference material or by spiking the test material with a known quantity of analyte) then a scheme could be technically viable with a single participant.

As far as financial viability is concerned, the question of minimum numbers is linked in with the fees charged. The VAM survey showed that the latter fall in the range of £80 to £2000 per annum, with a typical value being £300 per annum for

---

[†] However, it must be said that even two laboratories comparing results would be preferable to no intercomparison.

one set of tests in a scheme. The precise combination of minimum participant numbers and fees charged to enable a scheme to be financially viable will need to be considered on a scheme-by-scheme basis. As a general guide, organisers of new schemes should perhaps aim for a membership of at least 50 participants if the proposed scheme is to be entirely self-financed.

4.8 Costs of Scheme Organisation and Setting Fees

(a) How costs arise

In the operation of a proficiency testing scheme costs arise from the use of the resources identified in Section 4.4, namely:

- staffing, including specialist expertise;
- consumables;
- specialised equipment;
- accommodation.

Each scheme organiser will need to evaluate in detail the contribution each of these make to the running costs of their particular scheme.

However, for making a preliminary assessment of likely running costs, the following findings of the survey of UK schemes may be useful to organisers considering establishing a new scheme.

- A 'typical' PT scheme with 75 participants, covering 15 different tests (matrix/analyte combinations) and distributing materials on a monthly basis is likely to require a staffing level of 1 person year.
- Approximately one-third of the effort is likely to be concerned with the preparation and evaluation of the distributed materials, one-third with data handling tasks and one-third with participant liaison and management tasks.
- Consumable commodities take a number of forms (*e.g.* chemicals, test materials, packaging, postage, paper, telephone, photocopying, heating/ lighting, *etc.*). A typical annual cost for consumables might be £15k.
- Typically 20–50 m^2 each could be required for laboratory and office accommodation.

(b) Recovering costs from fees

If organisers costs are to be recovered from fees paid by participants, organisers must consider three main issues:

(i)   What is a realistic fee to charge so that participants will not be deterred from joining the scheme?
(ii) What is the minimum number of participants required?
(iii) What is the total number of potential participants?

It is impossible to give specific guidance on these issues, as the requirements of a specific scheme will depend on its own particular circumstances. However, the VAM survey suggested that a minimum participant number of at least 50 will apply and an acceptable annual participation fee (for one test within a given scheme) is likely to be about £300 (September 1992 figures).

4.9 Liaison with Potential Participants

It is difficult to recommend any formal mechanisms for identifying potential participants, but the survey showed that existing UK schemes rely on a number of approaches:

- brochures;
- lectures/presentations;
- advertisements in laboratory/technical/trade magazines;
- technical papers in scientific journals;
- contacts through professional bodies;
- personal knowledge of the measurement sector concerned;
- referrals from other participants;
- referrals from accreditation bodies *etc.*

Once identified, potential participants should be provided with a brief information pack outlining the essential features of the schemes operation, *i.e.*:

- scheme objectives;
- tests covered and types of test materials distributed;
- frequency of test material distribution;
- procedures used to assess and report participant performance;
- any ancillary technical advisory services offered;
- number of participants;
- fees charged.

All material used to publicise a scheme should contain a named contact point, address, telephone and fax numbers to facilitate enquiries from potential participants.

4.10 Advisory Committee

The practical day-to-day tasks involved in running a PT scheme are the responsibility of specific individuals or organisations. However, there are wider policy aspects of scheme management that need to be addressed. Such aspects will require a range of expertise and accordingly a steering committee should be formed to direct or advise the organiser on operational procedures. Many existing schemes adopt this approach and the VAM survey (Appendix 1) showed that the typical steering committee provided guidance on the topics covered in Box 4.4. The composition of the steering committee varies among schemes, depending on the issues to be addressed and the complexity of the scheme. Among the schemes surveyed, the number of steering committee members varied from four to sixteen, usually with no fixed period of service for members. Steering committees typically meet three times a year. The membership of a steering committee would usually include some or all of: scheme organisers; sub-contractors; professional bodies; trade associations; accreditation bodies; experienced analysts; specialist advisors (*e.g.* statistical experts).

Box 4.4 *Topics on which steering committees of proficiency testing schemes provide guidance*

- the tests to be covered by the scheme;
- ensuring the suitability of the test materials distributed;
- defining criteria for acceptable participant performance;
- ensuring data handling procedures are adequate;
- the format to be adopted for reporting participants' performance;
- dealing with complaints from participants;
- mechanisms for providing technical advice to poor performers;
- any other operational issues that arise.

The survey showed that in only a few schemes are participants' views specifically represented on steering committees. This is an aspect that organisers of new schemes should seek to address, although it is recognised that there may be difficulties in ensuring that the participant members are representative, and acceptable to the participants as independent of the organisers. It is often difficult for such representatives adequately to seek the views of other participants.

Similar considerations need to be given to securing the representation of end-users (*i.e.* the customers of test laboratories) on the scheme steering committee. Such representation is important since one of the prime objectives of proficiency testing is to ensure that analytical data are fit for the purposes of the ultimate end-user of the data. However, securing such representation may be difficult as, in some sectors, end-users are often unfamiliar with the concepts of data quality and fitness for purpose.

4.11 Higher Level Quality Body

A number of UK proficiency testing schemes (seven of the eighteen surveyed) have established links with other bodies in their field concerned with wider aspects of quality assurance. Such bodies are able to provide useful input to scheme management by advising on quality issues such as the selection of appropriate analytical methodology, mechanisms for assisting poor performers, the role of proficiency testing in the official licensing of analytical laboratories for particular tests and the use of proficiency testing in accreditation procedures.

Policy issues such as these may require wider consultation than allowed by a specialised steering board. Scheme organisers, therefore, should consider establishing links with appropriate bodies in their field of interest. By way of example, the work of various UK NEQAS steering committees is augmented by interaction with the National Quality Assurance Advisory Panel for Clinical Pathology, while the HSE's Committee on Fibre Measurement provides input to the steering committee responsible for the day-to-day operation of the RICE scheme.

4.12 Procedures for the Review of Scheme Operation

Organisers of PT schemes will need to review regularly at least two aspects of the operation of the scheme. At the conclusion of each round it will be necessary to check that no unforeseen aspects have arisen that throw doubt on the validity of the assigned value, the appropriateness of the target standard deviation or the suitability of the analytical procedures being used by participants. At longer intervals it is desirable to review the progress that the scheme is making towards achieving its stated objectives.

Round-by-round review

The results reported by the participants in each round should be examined for general deviations from expectations. Normally a unimodal distribution of results is expected, perhaps with heavy tails and some outliers. (There may be a distortion of results on the low tail if the analyte is present at a concentration not far removed from detection limits of the methods used – analysts have varying practices relating to the reporting of results at or below the detection limit.) A normal expectation would be for the central tendency of the raw results (*e.g.* the mode or a robust mean) to be close to the assigned value and their dispersion reasonably well described by a robust standard deviation somewhat smaller than the target value σ.

A multimodal distribution (*e.g.* Figure 4.1) is suggestive of the use among the participants of two or more discordant analytical methods. This should be investigated by requesting participants to disclose their methods to see whether there is a correlation between methods and results. If such a correlation is discovered, then the discordances should be publicised and appropriate action taken. In some instances such discrepancies may have other causes. They may result from a badly drafted analytical protocol or the use of a method outside the scope of its validation, *i.e.* on materials where interference effects are manifest.

If the distribution of results is not centred on the assigned value, (*e.g.* Figure 2.1) then either the assigned value is in error, the participants in general are not executing the analytical methods correctly or the methods themselves are biased. Any of these circumstances needs investigation by the organisers.

If the robust standard deviation of the results is greater than the target value (*e.g.* Figure 4.2), then the target value σ might be unrealistic in terms of the capabilities of readily-available analytical methods. This should be investigated by reference to statistics from any relevant collaborative trials or perhaps the Horwitz function.[22] In addition the criterion used to establish σ should be reviewed in

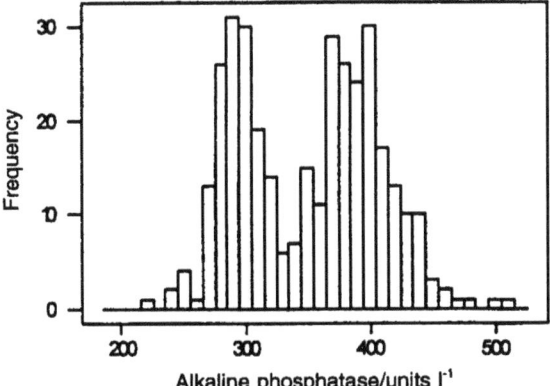

Figure 4.1   *Measurements of alkaline phosphatase in blood serum from a round of a proficiency test, showing the bimodal distribution resulting from the use of two different methods among the participants. (Data from UK NEQAS)*

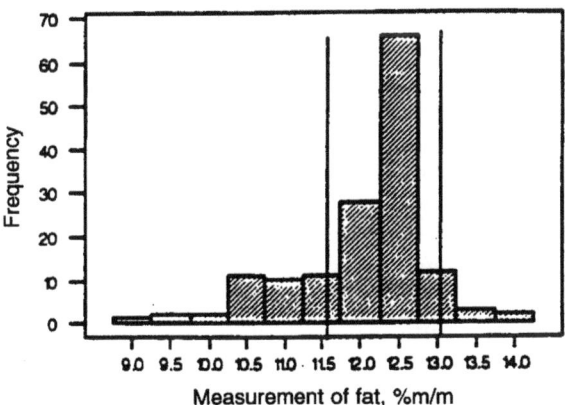

Figure 4.2   *Example of measurements (fat in a meat product) from a round of a proficiency test where the observed dispersion is greater than that specified by the target value for standard deviation σ. The vertical lines show the area bounded by $-2 < z < 2$, i.e. the 'satisfactory' results. (Data from FAPAS)*

terms of fitness for purpose (FFP) to see whether it is too stringent. The usual explanation of an over-large dispersion of results is that there are in use among the participants several different methods that are biased with respect to each other, but not to a degree that produces an overtly multimodal distribution of results (*e.g.* Figure 4.3). Again this may need investigation.

If there is a persistent discrepancy between a FFP criterion and general performance, then improved methodology is urgently called for. It is futile and potentially dangerous to produce data that are unfit for purpose.

Longer term reviews
Reviews of the progress of the scheme should be undertaken by the organisers and

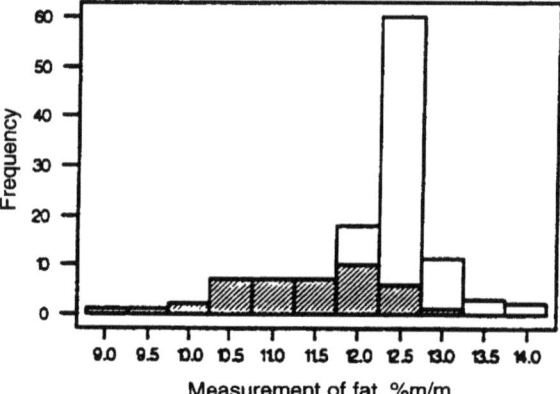

Figure 4.3   *Measurements of fat in a meat product in a round of a proficiency test. The apparently wide dispersion is explained when the results are distinguished by method. White bars show acid hydrolysis methods, hatched bars other methods. (Data from FAPAS)*

advisory committee after approximately one year or the passing of twelve rounds, whichever is the longer. The main purpose of the review is to determine whether the scheme is fulfilling its stated objectives. These objectives may change from the initiation of a scheme to its maturity. In the VAM context, a particularly important scheme objective is the promotion of improved data quality and organisers should review this aspect of their scheme's performance carefully. As a result of work carried out within the VAM project, two efficacy measures are proposed for monitoring the effectiveness of a scheme at securing improvements in performance by analytical laboratories.

The first efficacy measure EM_a is defined as: 'the percentage of participant laboratories producing results of acceptable quality'. As seen from Figures 4.4–4.7, this measure provides a simple means of assessing trends in participant performance over several rounds of a scheme. In general, the efficacy measure

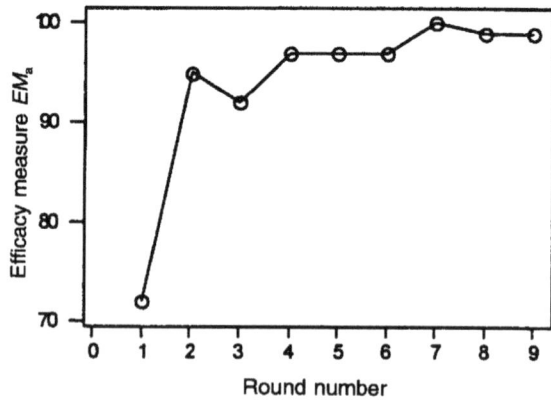

Figure 4.4   *Efficacy measure EM_a for laboratories undertaking asbestos fibre counting. (Data from RICE)*

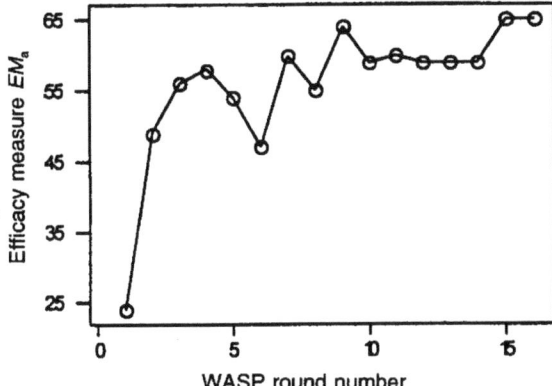

Figure 4.5 *Efficacy measure EM$_a$. The analyte was toluene absorbed onto Tenax. (Data from WASP)*

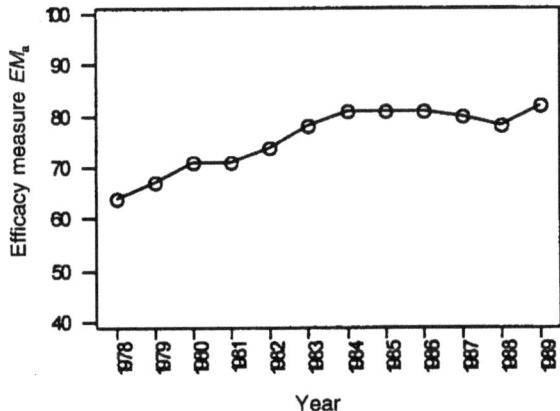

Figure 4.6 *Efficacy measure EM$_a$. for laboratories undertaking analysis of a wide range of analytes. (Data from US EPA Water Supply Studies)*

tends to improve over the first few rounds and then stabilise at some 'plateau' value. The data illustrated in Figures 4.4–4.7 are taken from the RICE scheme for asbestos fibre counting,[23] the WASP scheme for toxic substances in the workplace air,[24,25] the water supply studies of the US Environmental Protection Agency[26] and the cadmium in blood interlaboratory comparison programme of the Centre de Toxicologie du Québec respectively.[27]

This efficacy measure is recommended to scheme organisers as it gives a ready, if somewhat simplistic, picture of the general quality of data in the test sector concerned. However, when reviewing performance trends a number of factors should be borne in mind. Thus although there is ample evidence of data quality in PT schemes improving over time, the extent to which these improvements can be attributed entirely to the scheme is less clear. The analytical and quality assurance

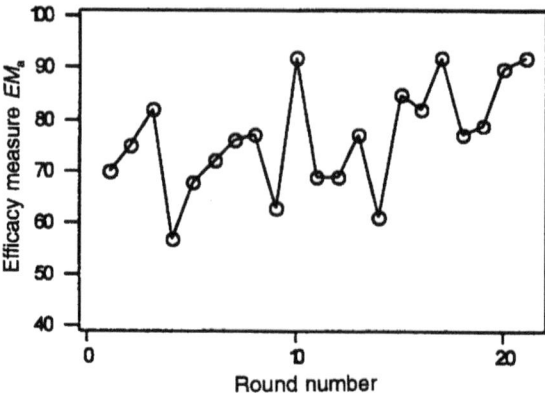

Figure 4.7 *Efficacy measure EM_a for laboratories undertaking the determination of cadmium in blood. (Data from Centre de Toxicologie du Québec)*

procedures used by test laboratories are rarely static and improvements in these matters will have a positive effect on a laboratory's performance in a PT scheme. Whether such improvements would have been made in the absence of the scheme is open to conjecture, but clearly an important role of proficiency testing is to make analysts aware of those measurement areas where such changes are needed. The efficacy of a PT scheme then depends in part on the readiness and ability of participating laboratories to improve their procedures and on the role the organisers adopt to encourage participants to make these improvements.

Interpreting the performance trends observed over several rounds of a scheme may be difficult, since the operational features of schemes may change from round to round. For example, the types of test material may vary as may the criteria used to assess performance. The number of participants is likely to fluctuate with time as newcomers join and existing participants drop out. Such changes will have an effect on the value of EM_a and this needs to be borne in mind when EM_a is used to assess the efficacy of a scheme.

A further efficacy measure is recommended and is defined as:

$$EM_b = \sigma / s_{rob}$$

where σ is the normal target value for standard deviation, and s_{rob} is the observed interlaboratory robust standard deviation. This quantity evaluates a scheme's ability to draw the spread in participants' results towards the chosen value of σ. For a scheme to be effective in this regard $EM_b \geq 1$. The measure is applied to the later rounds of a scheme, rather than the early rounds, so that the scheme is 'stabilised' (Figure 4.8) and rapid changes in participant performance are not occurring, allowing a reliable indication of scheme efficacy to be obtained. It is recommended that all scheme organisers should include the calculation of this measure in their long-term reviews of scheme operation.

The level at which a scheme stabilises is of interest. Ideally it should be such that the proportion of satisfactory participants is high ($EM_a > 95\%$) and that the robust standard deviation of results is slightly lower than the target value ($EM_b \geq 1$). Such a

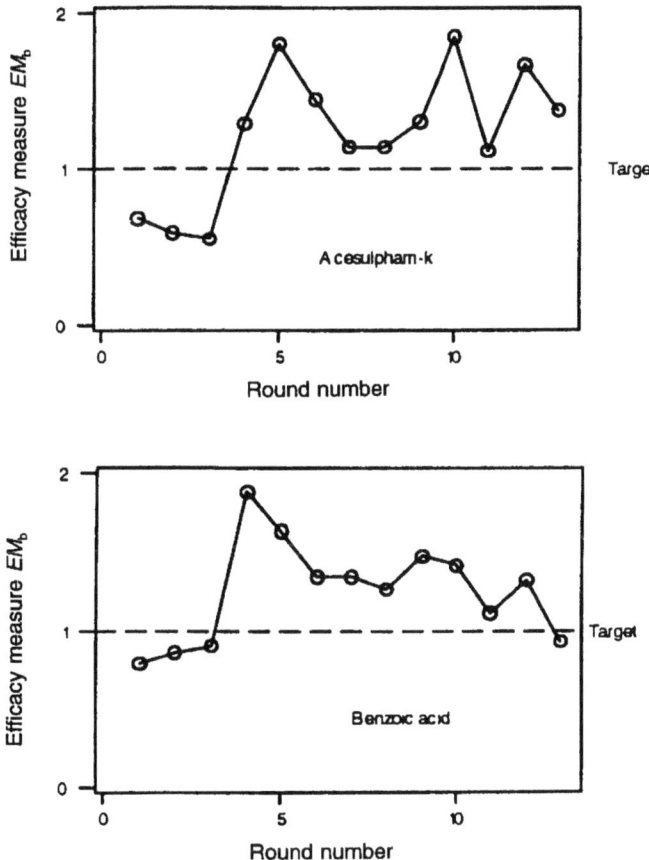

Figure 4.8 *Examples of efficacy measure EMb showing ideal behaviour. In the first few rounds the measure is below unity showing that (outliers aside) the laboratories as a whole were not performing to the standard implied by the target value for standard deviation. Subsequently the performance improves and either plateaus at just above unity (top) or increases sharply and then drifts down towards unity (bottom). The analytes were determined in soft drinks. (Data from FAPAS)*

condition would indicate that the majority of participants were performing satisfactorily. If that is not the case the organisers should consider possible actions to bring about a general improvement in performance as well as the measures specifically addressed to individual examples of unsatisfactory performance.

4.13 Certification of Schemes to External Quality Standards

As proficiency testing schemes evaluate the competence of participating laboratories, it is important that the operational procedures of the schemes themselves are evaluated, preferably by an independent body. As one means of achieving this, schemes could secure certification of their operational procedures to an appropriate quality system standard, such as the ISO 9000 series.[18]

For such an approach to be feasible the scheme's operational procedures must be documented in a quality manual. The manual then provides the basis for an independent audit of the scheme's operation. The International Harmonized Protocol[7] suggests a list of aspects of scheme operation that should be addressed in the scheme's quality manual. These are shown in Box 4.5.

It is seen from this list that many of the activities involved in the operation of a PT scheme are different from those undertaken in a laboratory purely concerned with making analytical measurements. Accordingly, the certification of a proficiency testing scheme to an external quality standard requires additional approaches to be developed. Because of this, work is currently (June 1996) underway at the Laboratory of the Government Chemist, as part of the VAM Initiative, to provide guidance on the documentation of a PT scheme's quality system. Follow-up work will address the issue of how such a quality system should be audited.

Box 4.5 *Topics to be included in a quality manual for a proficiency testing scheme*

Quality policy;
Organisation of agency;
Staff, including responsibilities;
Documentation control;
Audit and review procedures;
Aims, scope, statistical design and format (including frequency) of proficiency testing
    programmes;
Procedures:

- test material preparation and stability
- test material homogeneity
- assigned value
- equipment
- suppliers
- logistics (*e.g.* sample mailing)
- analysis of data

Preparation and issue of report;
Action and feedback to participants when required;
Documentation of records for each programme;
Complaints handling procedure;
Policies on confidentiality and ethical considerations;
Computing information, including maintenance of hardware and software;
Safety and other environmental factors;
Sub-contracting;
Fees for participation;
Scope of availability of program to others.

4.14 Ethical Issues

The organisers or the steering committee of a proficiency testing scheme may occasionally have to deal with issues of an ethical nature. The survey of UK scheme organisers showed that such issues do not arise frequently, but when they do the following are examples of the matters to be resolved.

Collusion between participants before submission of their results or fabrication of results by individual participants is not a widespread problem as far as the great majority of participants is concerned. However, occasional instances have been encountered and scheme organisers need to be aware of the possibility, as the integrity of the scheme could be threatened. Methods have been devised[8] to combat collusion. For example, the organiser could distribute in a round two very similar test materials that differ slightly in analyte concentration, so that half the participants (selected at random) receive one material and the remainder receive the second. Disclosure of this strategy by a scheme organiser, even as a possibility, would strongly deter collusion. Accreditation agencies have a role to play in detecting instances of collusion and falsification when they audit a laboratory's records of participation in a PT scheme. Such procedures should also reveal instances where a laboratory is applying special effort to the analysis of PT test materials over and above that it applies to its routine samples.

The use of performance scores by a participant laboratory to advertise its services to potential customers is an issue that scheme organisers may wish to consider, as the selective quotation of performance scores can give a misleading impression. Use of a combination score could also be misleading if it conceals a persistent inability to perform satisfactorily in a particular test.

The situation may arise where the organiser's test laboratory also participates in the scheme. In such circumstances it is important that procedures are in place to ensure that the test laboratory is treated identically to other participants and does not become aware of the assigned value of the test material or any special analytical difficulties associated with the material.

CHAPTER 5

The Organisation of Proficiency Testing Schemes – Technical Aspects

5.1 Overview

The detailed operational features of proficiency testing schemes inevitably vary from scheme to scheme, depending on such factors as the tests to be covered, the number of participants and the objectives of a scheme. However, there are several common features and a typical framework for an effective PT scheme can be identified. New schemes should be built around such a framework.

Box 5.1 shows, in flow chart form, the core practical activities required to operate a round of a PT scheme. Aspects of these activities are discussed in more detail in Sections 5.2–5.10. Scheme organisers will also need to undertake reviews of the operational effectiveness of the scheme and this aspect is discussed in Section 4.12.

5.2 Essential Properties of Proficiency Test Materials

Good quality materials are an essential factor if a PT scheme is to be effective. There are three important qualities which any proficiency test material must possess: the bulk material must be homogeneous, be stable at least within the timescale of a given round of a scheme and be similar in nature to the test materials analysed as routine by participants.

5.2.1 Homogeneity

Clearly, if all the participants within a scheme are to receive effectively identical portions of the test material, then the bulk material needs to be sufficiently close to homogeneous. However, only true solutions can be homogeneous in the strict sense, *i.e.* down to molecular level. In reality many test materials are multiphase (*e.g.* foodstuffs, soils) and cannot be made homogeneous in the strict sense. For the practical purposes of a proficiency test, a multiphase material has to be homogenised to a degree that residual differences between the compositions of the

Box 5.1 *Practical activities involved in the operation of a PT scheme*

CONSULT STEERING COMMITTEE
Identify test materials, analytes, analyte concentrations, the criteria for satisfactory performance (*e.g.* target standard deviation) and the means of establishing the assigned values of the test materials.

PREPARE/OBTAIN TEST MATERIALS IN APPROPRIATE QUANTITY

ESTABLISH SUITABILITY OF TEST MATERIALS
Verify homogeneity, stability and analyte concentrations are suitable. Establish assigned value by expert analysis if appropriate.

DISTRIBUTE TEST MATERIALS AND INSTRUCTIONS TO PARTICIPANTS
Advise participants of analyses required, date for return of results and any special features of the materials/tests involved.

RECEIVE RESULTS FROM PARTICIPANTS
Enter results into computer program, enter information on methods used by participants if appropriate.

EVALUATE PARTICIPANTS' RESULTS
Review spread in results – if unexpectedly large re-check homogeneity data. If appropriate establish assigned value from the consensus of participants results. Calculate participants performance scores (z-scores).

NOTIFY PARTICIPANTS OF THEIR PERFORMANCE SCORES
Prepare written report on the outcome of the proficiency test, detailing all essential information; distribute copies to each participant before the next round commences. Provide technical advice where necessary.

REVIEW THE OUTCOME OF THE ROUND
Check the distribution of data (is it uni- or multi-modal; is the robust standard deviation greater than the target standard deviation?) Check whether the participant consensus is centred on the assigned value established by expert consensus. Identify any tests causing significant problems for participants.   Report findings to the Steering Board.

CONSULT STEERING COMMITTEE AND INITIATE NEXT ROUND

samples distributed will contribute virtually nothing to the variability of the participants' results. This condition is called 'sufficient homogeneity'.

Once a bulk material has been fully homogenised and dispensed into units suitable for distribution, the homogeneity should be checked by selecting, at random, a given number of the units and analysing them for all of the analytes of interest. Where the results show the material to be not sufficiently homogeneous, the material must be re-processed or an alternative material selected. The only other approach would be to relax (*i.e.* make larger) the target standard deviation for that particular material, to take account of the variance that the material will contribute to individual participants' results. Except in minor excursions from sufficient homogeneity, such a practice would destroy the utility of the proficiency test. Moreover, participants would not have confidence in a scheme in which the distributed materials were not sufficiently homogeneous. The homogeneity of test materials is discussed in more detail in Section 5.5.

5.2.2 Stability

It is important that the distributed material(s) in a given round of a scheme are stable during the period between their preparation and their analysis by participants. Ideally, they should be stable for some time after the return of the results by participants in case there are any queries and problems that need to be addressed. Stability should be checked prior to the distribution of material by the PT scheme organisers. Stability checks may be carried out by accelerated testing at elevated temperatures to reduce the time involved in obtaining sufficient data. Alternatively, stability data can be obtained from the manufacturers of materials or through previously published data. Stability testing is discussed further in Section 5.6.

5.2.3 Similarity to Routine Test Materials

One of the important objectives of proficiency testing is to address routine operations, and therefore it is important that the proficiency test materials resemble routine test materials as closely as possible. In practice it is often difficult, if not impossible, to prepare PT materials that are of exactly the same type as routine test materials because of, for example, instability or problems of availability. In such circumstances the following guidance may be useful. The matrix does not necessarily have to be identical but should at least be generally similar to that of routine test materials. PT scheme organisers should also consider the needs of participants in terms of analyte/matrix combinations. Furthermore, the analyte concentrations in the test material matrix need to be of a similar magnitude to those encountered in routine samples.

A proficiency testing scheme organiser needs to be able to provide evidence to participants and other interested parties that the three requirements described above are fulfilled. If they are not, then a PT scheme is of limited use in assessing the analytical performance of laboratories in a given field.

5.3 Preparation of Proficiency Test Materials

Preparation of test material(s) typically occupies about 35% of the total effort needed to operate a PT scheme (Section 4.4). The following aspects should be considered.

Significant quantities of test material are likely to be required for each round of a scheme. Clearly, the amounts involved depend on numbers of participants and quantities required to carry out a given set of analyses. As a minimum, organisers need to ensure that sufficient material is prepared to enable homogeneity/stability checks to be carried out, for distributions to be made to all participants and for some to be kept in reserve should any queries arise. Organisers should also consider the preparation of a greater quantity of material than is needed for immediate use in a particular round. Such surplus test material is a valuable resource, both technically and economically, in that it may be used and marketed as a reference material or a quality control material.

'Natural' materials are preferred, *i.e.* materials that are obtained from the 'field', such as a food from a production line, a soil from a contaminated site or human serum samples from a donor bank. These are the best approximations to routine test materials, with the analyte in a 'native' state. However, it may be difficult to obtain such materials in sufficient quantity or such materials may not be stable for a long enough time period. In these instances there are a number of other options which the scheme organiser can consider.

A matrix can be spiked with the analytes if natural matrix materials cannot be obtained. However, although this option overcomes preparation problems, it is not ideal because the speciation or partition of the analyte in the spike may differ from that of the native analyte in untreated materials. Furthermore, the concentration of the native analyte in the matrix must be demonstrably zero or accurately known.

Substitute test materials may be used: for example, bovine serum may be used to simulate human serum, where the real matrix is difficult to obtain in suitable quantities. Synthetic test materials can also be employed. For example, pure beef fat may be used as a sample matrix to resemble for practical purposes the complex matrices of fat-containing foodstuffs (sausages, meat pies, *etc.*) that would be difficult to prepare homogeneously on a large-scale. However, in those circumstances where a substitute test material or a synthetic test material is normally used, scheme organisers should endeavour occasionally to distribute a real test material matrix, to check that participant performance on the substitute or simulated materials is an accurate reflection of their performance on a real matrix.

Problems of stability sometimes can be overcome by physically or chemically modifying the test material (*e.g.* by freeze drying or addition of a preservative). If that is done, it is important that the organisers check the effect of such modifications on the matrices concerned. Occasionally, such modified matrices may behave differently from natural matrices when analytical procedures designed for the latter are applied. For example, it may be necessary to issue instructions that freeze dried matrices are reconstituted as appropriate before analysis.

If the quantity of a test material required by participants for a given analysis is large, it may be impractical to distribute in its normal state. For example, in the analysis of water for trace contaminants one or two litres may be required. In such cases, a small volume of concentrated material could be distributed for subsequent dilution by the participants in their own laboratories. Again, organisers must give clear instructions on the dilution procedures to be used.

Where scheme organisers do not have the required facilities, the preparation of test materials can be carried out by a sub-contractor. In such cases, the scheme organiser must be satisfied that the sub-contractor is able to prepare the materials to the required quality and actually does so.

5.4 Numbers and Nature of Test Materials

For a scheme to be operated cost-effectively an organiser should aim to distribute in each round the minimum number of test materials that is consistent with providing an adequate assessment of laboratory performance. For example, in a study of data provided by the FMBRA and NACS PT schemes, the authors[28] have found that where several similar materials are analysed for the same analyte (*e.g.* different animal feedstuffs analysed for nitrogen) in one round, a laboratory's proficiency on one material correlates with that on the other materials. Possibly, one material may be sufficient to provide an adequate indication of proficiency in that test. An important function of the scheme's steering committee will be to establish what minimum range of test material matrices are genuinely required to provide an appropriate measure of laboratory competence in a given test.

A proficiency test material should potentially contain any or all the analytes of interest. However, in certain cases proficiency in the determination of one analyte may correlate with proficiency for other chemically similar analytes and therefore it may not be necessary to include all analytes in a test material, just a representative of the class. For example, in the case of organochlorine (OC) pesticides in food, it is more realistic to include only a few of the complete range of OC pesticides routinely sought. However, before such decisions are made the scheme steering committee must secure the appropriate experimental evidence to support such limited testing as a basis for inferring proficiency on a wide range of analytes. This is a complex issue because the *identity* of a pesticide within a class also needs to be established before the analysis can be regarded as satisfactory.

5.5 Homogeneity of Test Materials

Only true solutions are homogeneous in the strict sense, *i.e.* at a molecular level, and some proficiency test materials are indeed solutions. However, others are inherently heterogeneous, and scheme organisers must consider the benefits and disadvantages of complete (or as complete as possible) homogenisation. It may be a relevant part of the test for participants to overcome natural heterogeneity, for example, by 'homogenising' the contents of a can of meat before taking a portion for analysis. Even so, where a bulk quantity of test material is sub-divided into individual portions for distribution to participants, it is essential that the bulk material is *sufficiently* homogeneous for the purpose of the proficiency test. Specifically, it is important that any variability between portions is of a negligible magnitude in relation to the permitted variability between participants' results (characterised by the target standard deviation for the test concerned). This condition is known as sufficient homogeneity. This is a critical issue because a lack of confidence in this area can destroy the credibility of a proficiency testing scheme.

The test material therefore must be evaluated for homogeneity in some way before it is distributed to participants. However, this evaluation need not always entail a formal homogeneity check, although this should always be done on any new type of test material with which the organisers have no prior experience. It has also been observed that materials expected to be sufficiently homogeneous have occasionally failed formal homogeneity tests, so vigilance is necessary. Organisers must be fully aware that even if one analyte is homogeneously distributed in the test material matrix it does not automatically follow that all other analytes will also be homogeneously distributed. Specific evidence is required for each analyte. Because the cost of formal homogeneity testing makes a major contribution to the total cost of a proficiency test, consideration should be given to alternative procedures. It must be strongly emphasised, however, that the range of options available is small and limited to special circumstances.

The VAM survey of UK scheme organisers showed that where prior experience is available, the use of rigorous homogenisation procedures (*e.g.* blending, mixing, grinding, dissolution, *etc.*) known regularly to have produced a uniform bulk material may reasonably be presumed to be effective for similar materials prepared subsequently. However, in the light of the caution expressed above, such a presumption should always be verified by reviewing participants' subsequent results (see Section 4.12). An unusually large spread would indicate that there may be a problem with the homogeneity of that particular batch of test material or that participants experienced unusual problems with that particular test. Under such circumstances, a retrospective formal homogeneity check on the test material would need to be conducted by the scheme organiser and the participants advised of the outcome. However, that round of the scheme may well be rendered invalid if the material is found to be heterogeneous. Participants would have grounds for complaint if they were attributed poor performance scores as a result of being provided with a portion of non-homogeneous test material. This would have serious implications if the result were to be used for licensing or accreditation purposes in a commercial environment. Thus formal homogeneity testing, covering all analytes of interest, should be suspended only if the organiser has extensive experience of the test material concerned.

Where a formal homogeneity test is required the approach documented in the International Harmonized Protocol for Proficiency Testing of (Chemical) Analytical Laboratories[7] is recommended. This entails the following steps:

(a)   Divide the test material into the containers that will be distributed to participants.

(b)   Select a number $n \geq 10$ of the containers at random.

(c)   Separately homogenise the contents of each of the n selected containers and from each take two test portions.

(d) Use an appropriate method to analyse the $2n$ test portions in a random order under repeatability conditions (*i.e.* in a single run). (*Note:* the analytical method should have a repeatability standard deviation of lower than 0.3σ, where σ is the target standard deviation for the proficiency test concerned. Otherwise the test may fail to detect a significant proportion of cases of insufficient homogeneity.)

(e) Estimate the sampling variance, s_s^2, and the analytical repeatability variance, s_a^2, by one-way analysis of variance, without exclusion of outliers.

(f) Report values of $\bar{x}$, s_s, s_a, F and n.

(g) If F is less than the appropriate critical value for $p = 0.05$, then the material may be regarded as sufficiently homogeneous. In instances where F is greater than the critical value but s_s still less than 0.3σ, the material may still be regarded as sufficiently homogeneous. This follows because, even though it is demonstrably heterogeneous, variations between the samples of the material will not make a substantive contribution to the total variability of the participants' reported results.

Example. The following results are duplicate analyses (mg per 100 ml) for propan-1-ol conducted on ten random samples of a bulk alcoholic drink used in a proficiency test. The target value for standard deviation in this test was 2.2 mg per 100 ml.

Sample number	Analysis no. 1	Analysis no. 2
1	28.03	28.06
2	28.27	28.03
3	28.07	28.17
4	28.12	28.12
5	27.81	28.10
6	28.17	28.01
7	28.02	28.21
8	28.33	28.23
9	27.94	27.99
10	27.69	27.94

Analysis of variance on the data produced the following table of results:

Source of variation	DF	Sum of squares	Mean square	F	p
Samples	9	0.2952	0.0328	2.27	0.109
Analysis	10	0.1446	0.0145		
Total	19	0.4399			

There is no evidence of heterogeneity, as the value of F is not significant (as the critical value $F_{9,10} = 3.02$ at 95% confidence or, equivalently, the associated probability $p > 0.05$). We calculate that $s_a = \sqrt{0.0145} = 0.12$ and $s_s = \sqrt{(0.0328-0.0145)/2} = 0.096$. We can see from this that $s_a/\sigma = 0.05$, which is less than 0.3 (*i.e.* the precision of the method was good enough for the purposes of the homogeneity test). Also, $s_s/\sigma = 0.04$, which is less than 0.3, confirming that the material is sufficiently homogeneous for use in the proficiency test.

5.6 Stability of Test Materials

Test materials must be sufficiently stable to remain effectively unchanged in composition when exposed to the conditions of storage, distribution and the elapsed time from receipt to analysis of the material. Consequently, test materials will need to be evaluated for their physical/chemical stability by some appropriate means.

The survey of UK scheme organisers showed that few schemes have any formal procedures for assessing test material stability. Those that do employ accelerated testing at elevated temperatures to simulate the conditions a material will encounter during the proficiency testing cycle. However, it should be appreciated that such tests inevitably involve assumptions regarding the justifiability of extrapolating the results of accelerated testing to the real conditions of the PT exercise. Accelerated testing also requires a suitable datum against which the results can be judged. This might entail storing a further portion of the test material at $-20\,°C$, where it would be assumed that no degradation could occur. The two samples should then be analysed in replicate in a random order under repeatability conditions (in one run) and the results compared for a significant difference between the means by a two-sample statistical test.

Example. The following are simulated data for an accelerated stability test. The results on the samples subjected to the test were 56.9, 57.0, 55.8 and 57.6 and on the control samples 56.0, 57.0, 55.4 and 56.6. A two-sample t-test with pooled standard deviation gives the following statistics:

	N	Mean	Standard Deviation	Standard Error of Mean
Test	4	56.825	0.750	0.37
Control	4	56.250	0.700	0.35

The 95% confidence limits for the difference between the means is $(-0.68, 1.83)$ with six degrees of freedom, which shows no significant difference as it includes zero. (Alternatively, the observed value of t is 1.13 in comparison with the critical value of 2.45 for 95% confidence.) Hence there is no experimental evidence of significant instability.

Other means of evaluating test material stability include the organiser's prior experience, evidence from technical literature and the comparison of participants' results over a period of time following repeat distributions of identical specimens. However, where such ancillary information is not available, some type of formal stability assessment will be required.

Physical/chemical modification of a test material matrix (*e.g.* freeze-drying, addition of preservative, *etc.*) may be used to enhance the stability of the material. However, in such instances it is important that the effect of such modifications on the analytical methods applied to the test materials by participants is evaluated. If the behaviour of the modified material during analysis is significantly different from that of 'routine' test samples the value of the proficiency test is markedly reduced. It may be necessary to give special instructions to participants in respect of such matters; for example, participants might be requested to reconstitute freeze-dried serum, foodstuffs, *etc.* by addition of water in a prescribed manner.

5.7 Packaging and Dispatch of Test Materials

Test materials must be securely packaged and clearly addressed before dispatch, naming the responsible analyst in the recipient laboratory. The labelling should also indicate the nature of the contents (*i.e.* that they are proficiency test materials and not standard commercial items) and any relevant chemical hazards of the materials. The organiser's/sender's contact name, address, telephone and fax numbers should also be included on the labelling.

Those responsible for the dispatch of test materials should be aware of the relevant safety, postal and carriage regulations and ensure that these are complied with. Because of the hazardous nature (*e.g.* toxic, corrosive, flammable, *etc.*) or the chemical instability of some materials, specialist courier and packaging procedures may be required. If so, additional expense over and above normal postal charges will be incurred and should be allowed for in the fees to be charged to participants.

All documentation relating to the test round (special instructions, reporting pro-formas, *etc.*) should preferably be included with the test materials, rather than sent separately, if this is possible.

5.8 Frequency of Testing

An important feature of proficiency testing is the frequency with which test materials are distributed to participants. The survey of UK schemes showed that the frequencies currently used vary from as little as once a year to as much as once a week, with a typical frequency perhaps being every three months.

Frequency of testing will be decided by the scheme's steering committee, taking into account several factors, such as:

- resources required by the scheme organiser;
- availability of suitable test materials;
- operational costs incurred by the organiser;

- timescale involved in preparing performance reports, which should preferably be distributed before commencement of the following round;
- quality of participants' results in previous rounds;
- the frequency with which participant laboratories analyse routine samples;
- costs incurred by participating laboratories;
- regulatory/licensing requirements;
- accreditation requirements.

The survey of participants in schemes showed that participants were generally satisfied with the frequencies adopted by their particular schemes, which typically ranged from every month to every four months. Thus a frequency in this range could reasonably be expected to meet most participants' requirements without posing an undue burden on them.

As part of the VAM proficiency tesing project, the authors[28] have attempted to identify an optimum distribution frequency for proficiency testing based on historic data from seven different PT schemes. No clear picture emerged, but it appears that there are no strong grounds for believing that a frequency of three to four times a year is less effective in promoting improvements in data quality than a frequency of 24 times a year. The frequency of a PT scheme should never be less than twice a year, however, to preserve the regular nature of proficiency testing.

Finally, it should be noted that a distinction can be made between frequency of testing and frequency of test material distribution. It may be possible for scheme organisers to distribute several test materials in one batch to cover several successive rounds in a scheme. This offers the potential of saving on distribution costs, but due regard must be paid to the stability of the test materials involved. Also, there is the possibility that participants may analyse the materials in a single run while still reporting their results at the specified time intervals.

5.9 Instructions to Participants

When test materials are distributed to participants, instructions will need to be provided in relation to the analyses to be carried out. Such instructions will need to address some or all of the following:

- the analytes to be determined;
- reporting basis required (units *etc.*);
- method to be applied;
- pro-forma or similar for return of results and related information (*e.g.* details of method used);
- deadline for return of results;
- any abnormal hazards of the test material;
- any pretreatment to be performed on the test material (*e.g.* reconstitution of freeze-dried material, dilution of concentrated material, re-homogenisation of material).

In all of the above the overriding principle is that participants should be directed to apply exactly the same analytical procedures they use on routine samples to the PT materials. Only then does a PT assessment give a useful indication of the likely quality of a laboratory's routine data output.

5.10 Use of Computer Disks by Participants for Reporting Results

An alternative approach to the use of a pro-forma is to issue to participants a computer disk for entry and return of their results. This offers the advantage of reducing the transcription errors that are likely to occur when organisers transfer data from a pro-forma into a data processing program. A further advantage derives from the saving of staff time in the organisers facility, since manual data transfer is labour intensive.

The use of computer disks for return of participants' results was investigated as part of the VAM project on proficiency testing. A data entry program was devised and applied to the BAPS proficiency testing scheme, although the program is generic and applicable to any scheme. The results confirmed the benefits of adopting this approach.

5.11 Reporting Back to Participants

It is essential that reports of performance are distributed to all participants promptly, preferably before the next round commences, but not later than one month after completion of a round. If excessive delay is involved the benefits of the scheme will be diminished, as an undue period will elapse before the need for remedial action is brought to the participants' notice.

Reporting of results may be on an open basis, in which all participant laboratories are identified by name, or it may be on a confidential basis in which laboratories are identified by a code number only. In the latter instance it may be necessary to change code numbers from round-to-round in order to maintain full confidentiality. Currently in the UK nearly all schemes (16 out of the 18 surveyed) maintain participant confidentiality as it is believed to encourage participation. While the majority of participants in schemes (160 out of 210 replies) favour retaining confidentiality, most in fact would *not* be deterred from participating in their scheme if confidentiality were removed. A legitimate participant concern regarding removal of confidentiality that was revealed by the survey concerned the possible misinterpretation of performance scores by customers of laboratories. Organisers should therefore give clear guidance, in terms appropriate for non-experts, regarding the construction and meaning of the performance scoring system and the likelihood that even an experienced laboratory may occasionally receive a poor score.

The essential information to be reported back to participants is itemised in Box 5.2. Such information should be clearly presented. Graphical methods such as histograms or bar charts to illustrate the performance scores of all participants are strongly recommended. Appendix 3 provides examples of the reports issued to participants by a number of UK proficiency testing schemes.

As an important objective of proficiency testing is to improve analytical quality it is appropriate for reports to highlight instances of poor performance.

Box 5.2 *Essential information to be reported to participants*

- identification of the tests and the round number and date concerned;
- the nature of the test material and how it was prepared/obtained;
- results of the homogeneity assessments of the test materials;
- the assigned values and how they were established;
- the traceability and uncertainty of the assigned values;
- the target standard deviations and how they were established;
- the participant's raw results (so that the participant may check that the result submitted has been entered correctly by the organiser);
- the participant's performance scores calculated from the raw data;
- classification of performance as satisfactory, unsatisfactory, *etc.*;
- summary of the scores of *all* participants in the round.

5.12 Assessing Participant Performance

The organiser of a proficiency testing scheme has to report to the participants the results of each round of the scheme in a readily comprehensible manner. The recommended way of doing this[7] (see also Section 4.2) is based on converting the reported result x into a z-score given by:

$$z = (x - x_a)/\sigma$$

This involves establishing an assigned value x_a and a target value of standard deviation σ for each test in the round. Such a performance score adds value to proficiency testing by putting a result in a fitness for purpose perspective. The assigned value should, wherever possible, be accompanied by a statement of its uncertainty and traceability (see Section 5.13). The target value for standard deviation should be known in advance to the participants so that they know what standard of performance is required, and the value should be a consistent feature of the scheme so that meaningful comparisons of round-to-round performance can be made. As the concentration of the analyte is unknown to the participants before the test, it may be necessary for the target value to be expressed as a function of concentration, if the potential range of the analyte concentration is wide. All of the procedures used by the organiser for accomplishing the above-mentioned actions should be documented in a manual.

5.13 Establishing the Assigned Value

A number of measures are available to the organiser for establishing the assigned value for the analyte. These are in principle similar to the procedures required to assign a concentration value to a reference material.[29] However, the organiser of the proficiency test normally cannot expend as much effort as that required for certifying a reference material. Nevertheless, the assigned value is a critical feature of a proficiency test, and the use of an inappropriate value will reduce the value of the scheme.

The assigned value should always be the best practicable estimate of the true value of the concentration (or amount) of analyte in the test material. The use of a consensus of participants' results as an assigned value must be undertaken with caution. While it is effective in achieving consistency within the group of laboratories participating in the PT scheme, the results of this group might not be compatible with those of others within the same analytical sector. This would be an undesirable state of affairs because it prevents data from being fully transferable between institutions and hinders the mutual recognition of results. Methods used to establish assigned values are discussed below. Methods for establishing a consensus value from results are discussed separately (Section 5.14), as are methods of establishing uncertainties on assigned values (Section 5.15).

5.13.1 Use of Certified Reference Materials

A certified reference material makes an ideal test material for PT if the matrix is appropriate. The certified value can automatically be taken as the assigned value. By definition[30] there is already an uncertainty value ascribed to the certified value, and CRMs allow the traceability of analytical data to be established. At the present time, however, it is seldom that CRMs can fulfil this role in proficiency testing. They are unlikely to be available in sufficient quantities and the variety of materials certified is considerably smaller than the required range. Moreover the concept of a reference material is meaningful only when both analyte and matrix are stable over a long timescale, whereas many PT schemes deal with inherently less stable matrices such as foodstuffs and biomedical materials.

5.13.2 Consensus of Expert Laboratories

This assigned value is obtained by agreement between a number of expert laboratories that analyse the test material by the careful execution of recognised reference methods under good conditions. Preferably all of the experts should use the same validated method and traceable materials for calibration, because there is a better chance of achieving a consensus under these conditions. Failing this, the use among the laboratories of methods incorporating different physicochemical principles for the measurement process is advantageous. It is unlikely that problems such as matrix effects would affect the measurement process in the same way in the various methods and this helps to identify the magnitudes and likely sources of error. An estimate of the uncertainty can be obtained by consideration of the uncertainties reported by the experts and discrepancies between them.

5.13.3 Formulated or 'Synthetic' Test Materials

Formulation comprises the addition of a known amount or concentration of analyte to a base material containing no (or containing a small but well-characterised amount of) native analyte. The method is especially valuable when it is the *amount* of analyte added to individual test objects (such as air filters) that is the measurand. Formulation has the advantage that the pure analyte can usually be added in extremely accurate amounts by gravimetric or volumetric methods. Consequently there is usually no difficulty in establishing the traceability of the assigned value, and the uncertainty in the assigned value will be small and readily estimated.

There are some problems with the use of formulated test materials, however. Firstly, there may be problems in achieving sufficient homogeneity because of difficulties in mixing the analyte into the base material. This problem needs careful consideration when the matrix is a solid and when the analyte is added at trace levels. Secondly, the analyte is likely to be in a different chemical form from, or less strongly bonded to the matrix than, the native analyte in ordinary test materials that the proficiency test material is meant to represent. Unless this problem is demonstrably unimportant, representative materials (containing native analyte) are usually preferable in PT schemes.

5.13.4 Consensus of Participants

A value often used as the assigned value is the consensus of the results of all of the participants for that measurement. This value is clearly the easiest and cheapest to obtain. The method often gives a serviceable result when the analytical method is straightforward, easy to execute, its principles and potential drawbacks are well understood and it is well-characterised as a standard method. A consensus is the correct value to use when a single 'empirical method' is under consideration and many laboratories are involved, because by definition the true value necessarily falls within the uncertainty range of the consensus. When several distinct but recognised empirical methods are available for the determination of the same nominal analyte (*e.g.* 'fat' or 'fibre' in foodstuffs) special consideration has to be given to the assigned value to avoid discrepancies due simply to different choices of method among the participants. It is the task of the advisory committee of the particular scheme to resolve this difficulty in a manner that is appropriate to the analytical sector.

In some instances, however, the consensus of participant results is unsatisfactory as an assigned value. There is a theoretical problem that the traceability of the result cannot be established by conventional means, unless all of the participants use the same standard method and calibration standards. At a more practical level there are many documented examples of the consensus of a group of laboratories being seriously biased (*e.g.* Figure 2.1), and it is quite common to find that there is no consensus (in any useful meaning of the term) within a group of laboratories (*e.g.* Figure 3.4). These circumstances are not uncommon in trace analysis and where relatively new types of analysis are being undertaken. The consensus in such circumstances is not necessarily an unbiased estimate of the true value, and may serve to perpetuate the use of poor methodology. The adoption of the participant consensus should be made only after the advisory committee has given due consideration to these points and found that the risks are acceptable.

5.14 Procedure for Establishing a Consensus

For present purposes 'consensus' is taken to mean a broad agreement between laboratories and not exact unanimity. Obviously, at best, laboratories can only obtain agreement within bounds set by uncertainty of measurement. Moreover, in a large group there will usually be a small proportion of laboratories that produce discrepant results. The consensus is therefore taken to be the central tendency of the laboratories that are broadly in agreement. A consensus can therefore be obtained from a sample distribution of results that seems to be unimodal and, outliers aside, roughly symmetrical. It is futile to attempt to establish a consensus from a bimodal distribution or perhaps even a uniform distribution. Given these prerequisites, what methods are available to establish the numerical value of a consensus?

The classical estimators of central tendency are the mean, the median and the mode. The mode is an attractive estimator visually and intuitively because in a smoothed display of the ordered data the mode is the point of least slope (Figure 5.1) However, the mode is difficult to pinpoint accurately (although there are methods) and does not make full use of the information content of the data. The arithmetic mean is the minimum variance estimator, but is sensitive to the influence of outliers. The classical method of avoiding this influence is to apply outlier tests such as Dixon's Q or Grubbs's tests[31] and calculate the mean after exclusion of the outliers. However, outlier tests are seldom completely satisfactory, especially when many outliers may be present. The median is not sensitive to a reasonable proportion of outliers but, like the mode, does not make full use of the information content of the data.

A modern approach to the outlier problem is the use of 'robust statistics' in which the influence of outliers and heavy tails is downweighted. Robust statistics have been shown to be applicable to analytical data in a number of situations.[16]

There are many types of robust statistics. The median in fact is a robust estimate of central tendency in which the influence of heavy tails and outliers is negligible. A median-based estimate of standard deviation (the median of absolute deviations, or 'MAD') can also be calculated. Other robust statistics are based on 'trimmed' data[32] where a certain proportion of the results is excluded from the upper and lower tails of the data before calculation of the mean. The mean of the interquartile range is an example of this type. A more sophisticated approach is provided by Huber statistics[33] where the results (x) are weighted according to a

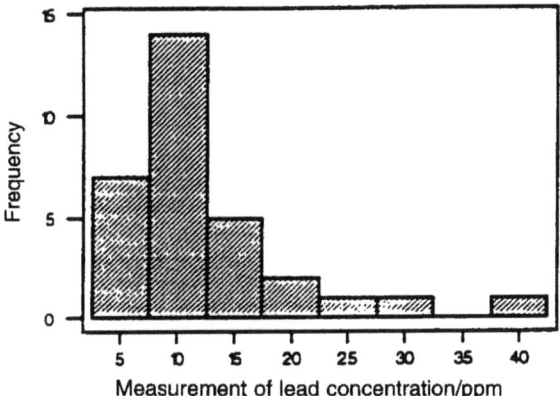

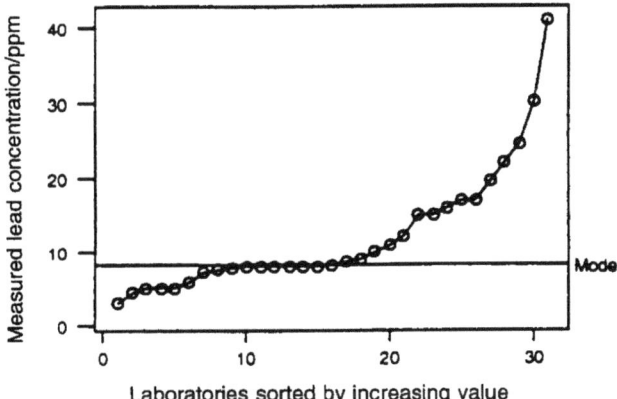

Figure 5.1   *Top: histogram of results from a round of a proficiency test for determining lead in a rock. The plot shows a mode at about 10 ppm, and the distribution is markedly skewed; bottom: x plot of the same data, showing the mode as a plateau region at 8.2 ppm. (Data from GeoPT)*

particular pattern before calculation of the statistics. Briefly, a value of x outside a range of $|x| > \hat{\mu} + k\hat{\sigma}$, $1 < k < 2$, are replaced by the value $\tilde{x} = \hat{\mu} + \text{sign}(x - \hat{\mu})k\sigma$. ($\hat{\mu}$ means 'estimate of μ'.) The robust statistics $\hat{\mu}_{\text{rob}}$ and $\hat{\sigma}_{\text{rob}}$ are estimated from the modified data by iteration from initial estimates of μ and σ. The constant k should be determined by the proportion of outliers; a value of 1.5 is normally used. A computer program is required for this calculation, and a suitable version is available.[16] The method is not suitable for small data sets.

Methods recommended for calculating the consensus value are therefore as follows:

(i)   For small ($n < 10$) data sets use the median.

(ii)   For larger data sets use one of the following (in preferred order): (a) the Huber robust mean; (b) the mean of the interquartile range; or (c) the median.

Examples. Here we consider two data sets, presented as an ordered series. Firstly there are determinations of a veterinary drug residue oxytetracycline (ppb) performed by five expert laboratories. This is a difficult determination and the values are somewhat variable: {105, 140, 182, 205, 213}. The results are shown in Figure 5.2. It is impossible to say whether the data represents a sample from a unimodal distribution, but the median of 182 seems an acceptable consensus.

The second example is the determination of fat (% m/m) in a foodstuff by a well-documented empirical method undertaken by 36 laboratories: {5.36, 5.50, 5.57, 5.59, 5.60, 5.65, 5.69, 5.70, 5.73, 5.75, 5.76, 5.78, 5.80, 5.80, 5.80, 5.82, 5.83, 5.84 5.84, 5.87, 5.90, 5.90, 5.92, 5.94, 5.95, 5.96, 5.98, 6.02, 6.03, 6.06, 6.08, 6.10, 6.13, 6.16, 6.18, 6.30}. The data are represented in Figure 5.3. In this instance there is a single mode between 5.8 and 5.9. The Huber robust mean is 5.86, the mean of the interquartile range is 5.86 and the median is 5.84. Any of these statistics could be used as a consensus.

Figure 5.2   *Dotplot of oxytetracycline data*

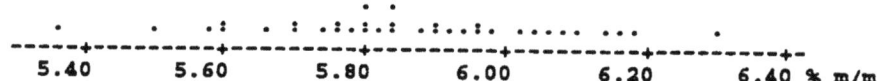

Figure 5.3   *Dotplot of fat data*

5.15 Uncertainty of the Assigned Value

Uncertainty is defined as 'a parameter associated with the result of a measurement that characterises the dispersion of the values that could reasonably be attributed to the measurand'.[34] In practical terms it is the interval on the measurement scale within which the true value lies with a high probability, when all sources of error have been taken into account. Analytical scientists are increasingly expected to supply an estimate of the uncertainty associated with their measurements as standard practice. Indeed a measurement cannot be properly interpreted unless it is accompanied at least implicitly by an uncertainty.

Recommended usage allows the expression of uncertainty either as a 'standard uncertainty' (u), expressed as a standard deviation, or an 'expanded uncertainty' (U), a quantity that defines an interval about the measurand that may be expected to encompass a large fraction of the distribution of values that could be attributed to the measurand. The former is more convenient for calculations, the latter for expressing the result as it emphasises the uncertainty as a range. The expanded uncertainty is obtained by multiplying the standard uncertainty by a 'coverage factor' k which in practice is typically in the range 2–3. A recommended method of expressing uncertainty u associated with a concentration c is $c\{U\}_k$, where U is the expanded uncertainty indicating a half-range. When k is omitted, a default value of 2 has to be assumed. If a default value of 2 is adopted U corresponds approximately to one half of the 95% confidence interval.

The uncertainty (u_a) on the assigned value of a test material in a PT scheme should be known to the organisers and ideally made available to the participants along with the assigned value in the report on the round. Should the uncertainty be of significant magnitude in relation to the target value for standard deviation (*i.e.* if $u_a > 0.3\sigma$) the usefulness of the PT is damaged because inappropriate z-scores may be reported to the participants. Attempts should be made to reduce the uncertainty of the assigned value to within this range.

When using a consensus we have to assume that the consensus is correct and outliers are incorrect. There is no alternative to this strategy in the absence of additional information. Of course subsequent information occasionally shows the assumption to have been unfounded. There are many documented instances of this (*e.g.* Figure 2.1). Advisory committees of PT schemes should therefore take every precaution to ensure that this assumption is justified in their particular scheme, and remain aware that the opposite may occasionally be true.

Several methods of estimating the uncertainty of the assigned value are potentially available. Where the test material is a CRM the uncertainty can be taken from the certificate. If the test material is prepared by formulation, the uncertainty is readily estimated by the combination of gravimetric and volumetric uncertainties in the standard manner.[35] When the assigned value is a consensus, however, the uncertainty is more difficult to estimate and at the present time no established practice has been recognised internationally. The following is therefore offered as a provisional guideline in estimating uncertainty of a consensus, a task that should be undertaken with the help of a statistical expert.

Consider the uncertainty u_a on a consensus from a moderate sized sample ($n > 20$) of laboratories. There will be a set of reported values x with (in principle) their individual standard uncertainties u thus:

$$x_1, x_2, \ldots\ldots x_i, \ldots\ldots x_n$$
$$u_1, u_2, \ldots\ldots u_i, \ldots\ldots u_n$$

(In fact participants do not report their uncertainties at the present time.) For a number of reasons it is unlikely that the participants' own uncertainties (even if available) will be used to calculate the uncertainty of the assigned value. It is therefore assumed that the true uncertainties within laboratories would be more accurately reflected simply by considering the interlaboratory variation. In that case the consensus could be estimated as $\hat{\mu}_{rob}$ from the x values by robust methods, with the uncertainty on the assigned value identified as $\hat{\sigma}_{rob} / \sqrt{n}$. These estimates would eliminate any undue influence of the outlying reported values of x on the uncertainty estimate. The value of $\hat{\sigma}_{rob}$ could be found by the Huber method described in Section 5.14 or by other suitable methods. Of course, the data would need to be inspected graphically beforehand, to make sure that there really was a consensus.

Example. Referring to the fat data in Section 5.14, we know that the data seem to be unimodal. Various robust methods give the following statistics:

	$\hat{\mu}$	$\hat{\sigma}$
Median/MAD	5.84	0.193
Interquartile range	5.86	0.204
Huber	5.86	0.207

The agreement between the various methods gives confidence in this approach. If we decide on the Huber statistics, the estimates give us an assigned value of 5.86 with a standard uncertainty of $0.207/\sqrt{36} = 0.035$. Thus we can express the assigned value as 5.86{0.07}, *i.e.* the true value falls in the range 5.79–5.93 with a probability of approximately 0.95.

If the number of reported values were smaller than specified in the foregoing, care would be needed in making the estimates. The Huber method is probably best avoided if there are less than ten observations, as may be the case if the consensus of expert laboratories is to be used. In such an instance the median of absolute deviations (MAD) method (see below) can be used to estimate the standard deviation. If the number of expert laboratories is very small, each case needs to be considered individually with the help of a statistical expert. It should be noted that the robust methods may fail if a large proportion of the results are identical.

Example. Referring to the oxytetracycline data in Section 5.14, the robust standard deviation can be estimated from the median of absolute deviations (MAD) from the median. The median of {105, 140, 182, 205, 213} is 182 and the respective absolute deviations are consequently:

$$\{77, 42, 0, 23, 31\}$$

The median of these absolute deviations is 31. The robust estimate of standard deviation is obtained by multiplying the MAD by a factor 1.48, giving:

$$\hat{\sigma}_{rob} = 31 \times 1.48 = 45.9$$

The estimated standard uncertainty on the assigned value is therefore:

$$45.9/\sqrt{5} \approx 21 \text{ ppb}$$

The assigned value and its uncertainty in recommended form is 182{42}.

5.16 Setting the Target Value for Standard Deviation

The target value for standard deviation defines the scale of the acceptable variation among laboratories in each particular test. It should be made known to the participants before the test, and the protocol of the PT scheme should document the basis on which the target value is established. Obviously the target value will vary with the concentration of the analyte (unless its possible range is small). The best method of expressing such variation is as a functional relationship such as, for instance, that described by Horwitz.[22] A method that attributes a small number of fixed standard deviations to particular ranges of analyte concentration is less realistic and gives rise to ambiguities in scoring at the boundaries of the ranges.

The same target value (or function) should be used over successive rounds of the proficiency test so that scores are comparable over time. The practice of using the standard deviation of the participants' results in the round as the target value is therefore to be deprecated, as this will vary from round-to-round. That could mask either the general improvement of participants round-by-round, or a general deterioration in the quality of data. It would enable organisers to identify discrepant laboratories with confidence, but would provide no information as to whether the discrepancy was important to the end-user of the data.

The best method of establishing the target value (or its functional relationship with concentration) is by reference to fitness for purpose (FFP). In other words, the precision specified by the target value should represent the maximum variation between laboratories that is consistent with routine laboratory data leading to correct decisions by the end-user. (This assumes, of course, that the laboratory's performance in the PT is representative of its routine performance.) At this time there are no generally accepted methods for quantifying the required performance in terms of fitness for purpose. However, it has recently been suggested that FFP is defined by minimising the sum of the costs of analysis and sampling plus the financial penalties associated with analytical errors. All of the terms in the sum are functions of the magnitudes of the uncertainties, so the method provides optimally cost-effective precisions as fit for purpose. However, no applications of this idea have been reported so far. In practice, therefore, the advisory committee of individual PT schemes will usually have to derive suitable criteria based on their professional experience of the use of data in that sector of the analytical community.

As a hypothetical example of such a criterion, it could be imagined that in a PT scheme dealing with the determination of lead in foodstuffs the target value was expressed as,

$$\sigma = 0.1 + 0.1c$$

where c is the concentration of the analyte in the test material and the results are expressed in ppm. Thus at a concentration of 1 ppm of lead:

$$\sigma = 0.1 + 0.1 \times 1 = 0.2$$

'Satisfactory' results (*i.e.* $|z| \leq 2$), would have to fall within the range 0.6–1.4 ppm if the assigned value were 1 ppm (*i.e.* $1 \pm 2 \times 0.2$). The corresponding range would be 7–11 ppm if the assigned value were 9 ppm (*i.e.* $9 \pm 2 \times 1.0$).

A different basis for the target value which is sometimes used in PT is to specify the precision that might reasonably be possible, given the current state of analytical technology. This criterion might be based, for example, on the reproducibility standard deviations obtained from a collaborative trial (method performance study), or on other aspects of the analytical method or laboratories. This approach is appropriate only if it is consistent with fitness for purpose. If such a criterion is inconsistent with fitness for purpose, the PT scheme will not be encouraging participants to achieve the standard of accuracy that is actually required for the job in hand. This could result either in the wasted expense of forcing laboratories to achieve an unnecessary degree of accuracy, or leading them to suppose that poor accuracy was in fact acceptable. Advisory committees should consider whether such 'reasonably possible' criteria are appropriate for their sector of the analytical community.

5.17 Performance Scores

Scoring is the method of converting a participant's raw result into a standard form that adds judgmental information about performance. The requirements for any method of scoring are that it should be: (i) simple and readily understandable; (ii) give rise to scores of comparable magnitude (for comparable performance) for different tests and different PT schemes; and (iii) universally applicable. To make a score readily understandable it should ideally be in the form of a standard statistic without the use of arbitrary scaling factors. Scaling has sometimes been introduced into PT scoring in order to avoid negative numbers and decimals, an unnecessary action within the scientific community. If the score is of comparable magnitude in all PT schemes, an analytical chemist would be able immediately to assess the significance of a score from any PT scheme. The method of scoring must be equally applicable to any proficiency test that gives rise to data expressed on an interval or ratio scale, which most measurements in analytical science fulfil. The z-score recommended here meets these criteria and is probably the most effective of the available possibilities (see also Sections 3.2 and 5.12).

The interpretation of a z-score is based on the properties of the standard normal deviate. If in fact the results from a group of participants, obtained in a round of a PT scheme for a particular test, correspond to a normal distribution with a mean equal to x_a and a standard deviation of σ (*i.e.* if they are perfectly fit for purpose), then the z-scores will be a sample from the standard normal distribution, *i.e.* with zero mean and unit variance symbolised as $N(0,1)$. About 5% of the results would fall outside the limits of ± 2, and only about 0.3% would fall outside ± 3. Of course these circumstances are rarely fulfilled. Actual distributions from PT schemes deviate from the normal, often by possessing 'heavy tails' (*i.e.* an unduly high proportion of results with $|z|$ falling between about 1.5 and 3) and/or a number of outliers. While the central tendency often is close to the assigned value, there is no guarantee that this will happen. Moreover, the actual dispersion of the results may differ appreciably from the target value: if it is larger then an undue proportion of z-scores absolutely higher than three will result. Often then there will be an apparent excess of absolutely high z-scores. This does not detract from the use of the z-score, so long as the target value of standard deviation is set according to criteria based on fitness for purpose. The high values are interpreted to mean that an unacceptable error has been produced by an individual laboratory. As a consequence, z-scores outside the range ± 3 indicate that an investigation of the cause and/or remedial action is necessary within the participant's laboratory. A value outside the range of ± 2 is a warning of potential problems.

 z-Scores can be combined in a number of ways that give an average or summary of performance over several tests. Combination scores that are statistically well founded are discussed elsewhere (Section 3.5). A combination score can be a convenient method of producing an overview of a participant's performance for review purposes. For example it may be desirable to say how a participant fared in the same test over a period of one year. This would be a legitimate use of a combined score, although it must be remembered that poor performance on one or more occasions may be masked by a combination score. While this situation is unlikely to arise by chance in a laboratory performing well, it is more likely in a laboratory with erratic performance. Such a score must be interpreted with due caution.

Sometimes a participant will want to have a score representing the performance averaged over tests on several materials in the same round. This is legitimate if the tests are all of the same nature, *e.g.* for the same analyte in a range of similar matrices. However, it is not legitimate to use such a score if the results of different tests (*i.e.* different analytes and/or matrices) are to be combined. The combined score could mask the fact that the participant was *always* below standard for a particular test.

Overall it is not clear that combination scores are very helpful. A graphical display of individual z-scores provides more information, and is less open to misinterpretation. Such a display could be used to look at trends within the results of a single test in a particular laboratory, or to compare the results of different tests in a round (Figure 5.4).

5.18 Classification of Performance

The primary purpose of PT is the detection of inaccuracy in a laboratory's results so that remedial activities can be instigated and their effects monitored. In this view the scheme is essentially for the benefit of the participants. Therefore the critical values of the z-score are regarded as decision limits for the implementation of remedial activity. However, there is an increasing tendency for accreditation and licensing agencies to use the results of proficiency tests for classification purposes. Some schemes produce a published list of participants regarded as satisfactory, and a presence on the list corresponds to a licence to conduct that type of analysis. If this is to take place then it is essential that the classification scheme is well founded, fully documented and well-known to the participants. It

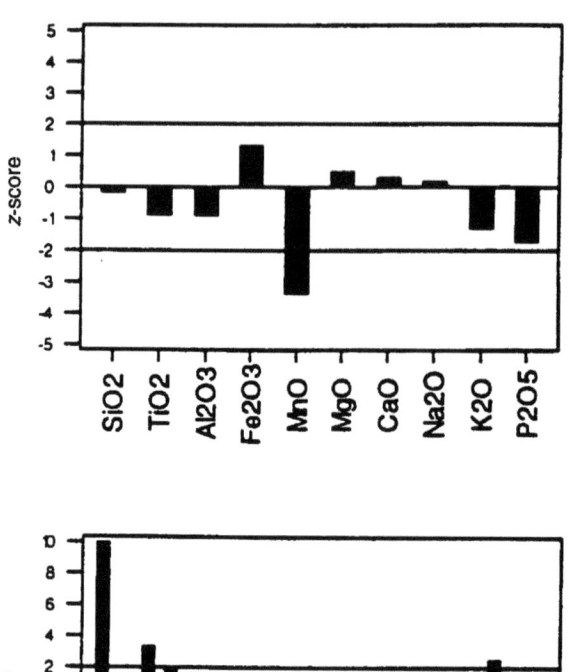

Figure 5.4 *Top: plot of z-scores obtained by different analyses in a round of the scheme. This is a convenient and informative way of summarising the data. (Data from GeoPT). Bottom: plot of z-scores from successive rounds of a PT scheme obtained by a participant for a single analyte*

must also be remembered that there is a fundamental limitation on all overt proficiency tests: they address the ability of participants to perform in proficiency tests – the extrapolation of test performance to everyday performance is an assumption for which there is no direct evidence, except where covert tests are also used.

Classification of a single z-score is straightforward. If the participant's result corresponded to the specification described by x_a and σ, there would be approximately 5% chance of the z-score falling outside the limits ±2, and 0.3% chance of falling outside ±3. Consequently we can classify single scores as follows, although it must be recognised that the limits set and the descriptors are arbitrary:

$$|z| \leq 2 \qquad \text{satisfactory}$$
$$2 \leq |z| \leq 3 \qquad \text{questionable}$$
$$|z| \geq 3 \qquad \text{unsatisfactory}$$

Classification of a laboratory as a whole is a more complex task, even in the simplest case. The organiser and advisory committee of a PT scheme have to take into account all of the circumstances surrounding the scheme, the participants and the end-users before devising the details of the classification protocol.

The simplest case to consider is where there is only a single test involved. This test is conducted on one material in successive rounds. Classifying the participant on the basis of each round may seem too rigorous, especially if accreditation or licensing is at stake. A more flexible system would be to classify on the basis of a running score (Section 3.5), probably an *RSZ*, extending over several rounds (Figure 5.5). The same criteria used for classifying a single z-score can also be used on the *RSZ*. However, a single large z-score in an early round could make the *RSZ* large long after the problem giving rise to the exceptional score had been eliminated. This 'memory effect' in *RSZ* (or any other combination score) could unfairly penalise a laboratory that had already taken effective remedial action. Therefore, for the purpose of calculating a running score, the individual scores could be restricted to a range such as ±4. A value outside this range would be replaced by the value of four with the appropriate sign. An alternative classification that would achieve much the same effect would be to regard a laboratory as satisfactory if it achieved an 'acceptable' score in (say) three out of the four previous rounds. An approach of this type is used in clinical proficiency testing in the USA[36] and in the UK's RICE scheme for asbestos fibre counting.[23]

Similar considerations apply in the case where the PT scheme involves the analysis of several similar materials for the same analyte in each round of the test. In this case the laboratory could be classified on the *RSZ* or other combination score for each round. However, since the test materials are likely to be analysed

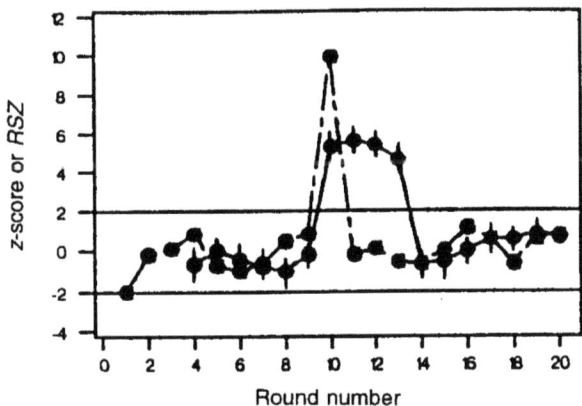

Figure 5.5   *Series of z-scores (open circles) with a single outlier in round 10, showing the memory effect in the corresponding RSZ value for the round and the preceding three rounds (solid circles)*

together under repeatability conditions, any bias present at the time of analysis is likely to affect all of the materials to the same extent. Hence classification on the basis of one round could be equally rigorous as in the simple case where only one test is made per round. Again it would seem preferable to classify on the basis of a running score with appropriate truncation of the individual scores, or to use the proportion of acceptable scores over several rounds as a criterion.

More difficult is the case where many analytes are determined during the course of a round. As previously noted, the *RSZ* on a single round could easily mask the fact that the laboratory was performing consistently but unacceptably for one particular analyte or a small proportion of the analytes. In such an instance classifying on the *RSZ* would certainly be misleading and probably detrimental. In fact any classification scheme that treated results from different analytes as equivalent would suffer from the same defect. The only remedy is to include in the scheme a feature that treats each analyte on a separate basis. An example of a rule of this type would be: "for the current round the *RSZ* over all analytes must be 'acceptable' and for each analyte the *z*-score must be 'acceptable' at least three times out of the current round and previous three rounds".

5.19 Ranking of Participants

Some scheme organisers and many participants favour the production of a list of laboratories at the end of each round of the PT, ranked according to the magnitude of the score achieved. This practice is not recommended for the following reasons. Firstly, the ranking must be executed on a single combination score covering many analytes, and it is clear (Section 3.5) that such a score can mask consistently unacceptable performance in one or more particular analytes. Secondly, the information content of a list (or diagram) of actual *z*-scores is much higher than a ranked list of participants. Finally, the rank obtained by a laboratory performing consistently well can be extremely variable even when derived from a combination score. This is a result of the substantial random component in the majority of analytical errors. As a consequence it is difficult to interpret the rank of a laboratory and at least equally difficult to compare the ranks obtained in successive rounds. Therefore the rank attributed to a laboratory would probably be misleading and this fact will not generally be appreciated even within the scientific community.

An often-cited reason for producing a ranked list of participants is that it is easy for a non-scientist to understand. However, such a list is just as easy to misunderstand, and even to misrepresent. It is better for organisers to avoid this situation by refusing to produce a ranked list.

5.20 Uncertainties of Measurement on Submitted Results

The widespread use of uncertainty of measurement in analytical science could ultimately mean that participants in proficiency tests may be expected to report to the organiser their estimates of concentration (or other measurement) accompanied by the respective uncertainties. This is a consequence of the ideal that participants there is no experience in proficiency testing schemes of handling participants' uncertainties.

When a laboratory submits a value x with standard combined uncertainty u_x we could construct two resultant scores, the ordinary z-score and an additional 'zeta-score', namely $\zeta = (x - x_\text{a})/\sqrt{(u_x^2+u_\text{a}^2)}$. The additional use of ζ would allow each laboratory to be scored on the basis of its own claimed uncertainty value, presumably fit for purpose in its own field of work, thus allowing laboratories with quite different functions to participate meaningfully in a single proficiency testing scheme.

A high value of ζ would penalise a participant for either an inaccurate result or an unduly small estimate of u_x. Unsatisfactory values of z and/or ζ would indicate problems with uncertainties to end-users of data. However, there remains the possibility that a participant could achieve a small absolute ζ-score by reporting a arbitrarily high value of u_x. It would be the responsibility of end-users and accreditation agencies alike to ensure that the value of u_x used by the laboratory was consistent with fitness of purpose, *i.e.* that the uncertainty of the result was sufficiently small to make the result useful.

CHAPTER 6

Participation in Proficiency Testing Schemes by Analytical Laboratories

6.1 Benefits of Proficiency Testing

The benefits of proficiency testing to participating laboratories can be wide ranging, although the nature and degree of the benefits depend largely on the type of participating laboratory and the particular scheme in which it is taking part. Proficiency testing provides some or all of the benefits shown in Box 6.1.

The most important benefit of proficiency testing is that it provides a regular assessment of a laboratory's performance, test samples being distributed typically between three and twelve times a year. It is also an external and independent means of assessment since the assigned values for the test materials are not disclosed to participants until after all results have been reported. A further benefit of proficiency testing is that it provides laboratories with an incentive to improve their performance. Practical experience has shown that over consecutive rounds of a scheme, the performance of participants improves. This is particularly true in the early stages of participation in a scheme (Section 4.12). It is also important to note that accreditation/certification bodies are increasingly requiring that a laboratory participates in a proficiency testing scheme (where one is available) before accredited status will be awarded (Section 7.7).

The direct comparison with peer laboratories that participation provides is a unique feature of proficiency testing schemes as no other quality assurance measure provides this. Laboratories can readily determine whether their results are in accord with the general level of competence in the measurement concerned. It will be appreciated, however, that agreement with the consensus is not necessarily indicative of trueness, as the consensus could be biased.

Scheme organisers will include in their steering committees analytical chemists who are experienced in the particular fields of analysis covered by their scheme and are therefore well placed to offer participants technical advice on any analytical problems that they may encounter. Experience has shown that such advice and feedback is effective in helping laboratories to improve their overall performance in the schemes.

Box 6.1. *Benefits from participation in proficiency testing*

- a regular, external and independent check on data quality;
- assistance in demonstrating quality and commitment to quality issues;
- motivation to improve performance;
- support for accreditation/certification to quality standards;
- comparison of performance with that of peers;
- assistance in the identification of measurement problems;
- feedback and technical advice from organisers (reports, newsletters, open meetings);
- assistance in the evaluation of methods and instrumentation;
- a particularly valuable method of QC where suitable reference materials are not available;
- assistance in training staff;
- assistance in the marketing of analytical services;
- savings in time/costs by reducing the need for repeat measurements;
- a guard against loss of reputation due to poor measurement;
- increased competitiveness.

Many schemes collect information from participants on the methods and instruments they have used for the analysis of the test materials. As different participants often use different analytical procedures for the same matrix/analyte combination, a presentation of performance scores in terms of the methods used enables laboratories to identify any particular method-based problems that may exist.

Although the use of reference materials is an important component of IQC, suitable reference materials are not always available and they may be expensive. Proficiency testing is particularly valuable as an alternative QC measure in such cases as it provides the analyst on a regular basis with test materials that are a good approximation to reference materials.

There are a number of commercial benefits that can accrue to laboratories as a result of their participation in proficiency testing schemes. They can use their participation in a scheme as a marketing tool for their analytical services as it is an effective demonstration of their commitment to quality. Furthermore, participation helps laboratories to identify measurement problems and they are therefore less likely to incur either financial loss or damage to their reputation as a result of providing poor data to their clients.

6.2 Costs of Proficiency Testing

Clearly, as well as deriving important benefits from proficiency testing, laboratories will also incur costs from participation in PT schemes.

The most easily identified and quantifiable cost is the fee which participants pay to the organisers of schemes. The VAM survey of UK PT schemes conducted in September 1992 (Appendix 2) showed that participants can expect to pay fees in the range of £80 to £2000 per annum for one group of tests within a particular scheme. When it is realised that, for a given test, (different) test materials are distributed between three and twelve times per year, and that each distribution often includes more than one material, it is readily apparent that the fees charged represent good value for money. Most PT schemes operate on a cost-recovery, non-profit making basis; where a profit is made it is usually small (approximately 25%) and is used to fund further development and improvement of the scheme.

Added to the fees, of course, are the costs of the analyses of the PT materials, arising from staff time, consumables and overheads. These costs clearly depend on the number of PT materials analysed, but where this number is small in comparison to the number of routine tests of the same type carried out, the cost of PT participation will be small relative to total analytical costs. However, it will be appreciated that if only few routine tests of the type covered by the proficiency test are carried out, then the costs of PT could form a significant part of analytical costs. In these circumstances, each laboratory will need to consider carefully its reasons for participation. Laboratories should appreciate that although the benefits listed in Box 6.1 are difficult to quantify in monetary terms, they are often of significant value and therefore compensate for the cost of participation.

6.3 Existing Schemes

Although a number of proficiency testing schemes in the fields of chemistry and microbiology are currently operating in the UK, a problem analytical laboratories sometimes experience is locating a PT scheme suitable for their requirements. Consequently, a register of current UK schemes has been published[10] by the Eurachem-UK Proficiency Testing Working Group and gives information to potential participants on the scope of each scheme and the organising body (see Appendix 1).

Compilations of proficiency testing schemes operating in other countries have also been produced, but not necessarily published in readily accessible literature. For example, the Federal Institute for Materials Research and Testing in Berlin has carried out a survey of PT schemes currently operating in Germany. Similar surveys have been made of schemes in France, the Netherlands and Eire. An international workshop on interlaboratory studies, organised by EURACHEM and held in the Netherlands in 1995, identified the need for better and more comprehensive information on PT schemes in Europe.

The American Association for Laboratory Accreditation (AALA) maintains a listing of PT schemes, participation in which is required of laboratories seeking accreditation from AALA. Also in the USA, the Association of Food and Drug Officials has published a list of proficiency testing schemes.[37] The National Association of Testing Authorities, Australia (NATA) produces a 'Proficiency Directory', which includes schemes from Europe and the USA, as well as schemes operating in Australia.

6.4 Practical Approach to be Adopted by Participants

When laboratories participate in proficiency testing schemes it is important that they fully appreciate the purpose of the exercise, which is to assess the quality of a laboratory's routine data. Satisfactory performance in a scheme is not an end in itself; it is of value only to the extent that it indicates satisfactory performance (or, alternatively, the need for remedial action) in a laboratory's day-to-day work.

Consequently, it is essential that the PT test materials are treated in exactly the same manner as the unknown samples routinely analysed by the laboratory in its day-to-day work. For example, the analyses should be carried out by the usual analysts (not specially selected or experienced analysts), using routine analytical methods, with the number of replicates carried out on the test sample being the same as for routine samples. Normal reporting and checking procedures should also be employed. Proficiency testing can only be considered to give a true picture of a laboratory's performance where these conditions are met. Where a laboratory deliberately modifies its routine procedures in the hope of obtaining a good performance score, the value of that score as a diagnostic tool to the laboratory itself as well as to other interested parties is greatly diminished. Laboratories should therefore observe the proper spirit of PT and analyse test materials in exactly the same manner as routine materials.

The great majority of laboratories participating in proficiency testing schemes do so in an honest and professional manner. However, a minority of participants may be tempted to fabricate results or to amend their results after discussion with other participants before reporting their data to the scheme organiser. Such actions are deprecated not only because they diminish the standing of analysts in general, as well as the reputation of the individual laboratory concerned, but because they destroy the value of the proficiency test for the particular laboratory. It should be noted that there are ways of organising proficiency tests that would strongly deter collusion among participants (Section 4.14).

As participation in proficiency testing schemes increasingly becomes a requirement of laboratory accreditation, participant laboratories should be aware that an evaluation of their approach to PT participation may form part of the accreditation audit and, by this mechanism, irregular procedures may be identified and perhaps made known to the organiser.

It is also important that laboratory managers train their staff in matters relating to PT, for example its role, the basis of performance assessment, the need to treat test samples as routine, *etc.*, so that the analysts involved tackle the exercise in the appropriate manner.

6.5 Internal Review of Performance

In order to ensure that appropriate lessons are learnt and that any corrective action necessary can be taken quickly, participants must establish mechanisms for formally reviewing their performance after each round of a scheme. The reviews should ideally involve not only the laboratory manager/quality manager, but also the analysts involved in testing PT samples and it is important that such reviews are undertaken before the next round of the scheme, wherever possible.

The exact format of the review will vary from one laboratory to another, but will include an evaluation of the laboratory's results in relation to the assigned value. Any significant deviations must be investigated and a z-score of outside the range ±3 will require immediate attention, as the probability of a satisfactory laboratory achieving such a score purely by chance (rather than by poor analysis) is small. A concurrent review of internal quality control data (*e.g.* the appropriate control chart) may indicate whether the poor PT performance is due to poor statistical control of the analysis (*i.e.* poor precision). If the latter is satisfactory, the poor result will be due to a bias (lack of trueness) in the measurement, which could arise in a number of ways (*e.g.* mistakes in procedures, incomplete extraction of analytes from sample matrix, interferences, inadequate instrument calibration, *etc.*). The effective integration of proficiency testing with other forms of analytical quality assurance (Section 2.4) that address these aspects of analytical measurements will enable the source of any poor performance to be identified more readily. Where the necessary ancillary information on data quality (precision and trueness) is not available, some experimental investigation will be required in order to identify the source of the problem.

Occasionally an unsatisfactory *z*-score may be due to the use of inappropriate methodology, rather than to poor execution of that methodology. For example, such problems may arise where empirical determinations (*e.g.* the fat content of foodstuffs) are involved, wherein the method defines the fat content. If a laboratory uses a different empirical method to that used to establish the assigned value, a poor *z*-score may result even though the method has been applied correctly.

If, after reviewing possible causes of poor performance, a participant is unable to identify the source of the problem, scheme organisers are often able to provide useful advice and guidance. The outcome of internal reviews of performance in a scheme and any resulting corrective action taken should be fully documented. This is of particular importance where participation in a PT scheme is recommended or required by accreditation and certification bodies.

6.6 Participant Input to Scheme Design

Proficiency testing schemes are set up to meet the needs of the analytical community involved and, accordingly, the participating laboratories themselves have a role to play in ensuring the schemes are meeting their needs. Scheme organisers therefore will be interested to receive comments and suggestions regarding, for example, the types of test included in the scheme, the nature of performance assessments, the clarity of reports, the use of the scheme to assess analytical methods or any other issues of interest to participant laboratories.

It should also be noted that many schemes will have a steering committee or similar body to provide guidance on the direction and management of the scheme. Participant representation on such a body is important although it must be recognised that effective participant representation will require a mechanism (*e.g.* via a professional body) for representatives to communicate with other participants.

Other formal avenues for participant input to scheme design and operation should include such activities as regular open meetings for scheme participants and questionnaires, *etc.* to establish participant views.

6.7 Formal Recognition of Participation

Although scheme organisers will provide each participating laboratory with a report on their performance after each round, such reports will be on a confidential basis, each laboratory being identified by a code number. These reports enable laboratories to evaluate their own performance and also to demonstrate their participation and performance in the scheme to third parties such as accreditation agencies, customers and regulatory authorities. As an additional recognition of satisfactory performance in a scheme, organisers could regularly compile (where it is agreed by all concerned or required by legislation) a listing by name of all laboratories that have participated and achieved an overall satisfactory performance in the scheme. The definition of overall satisfactory performance will vary depending on the purpose and design of the scheme and various approaches are discussed in Section 5.18. In general terms, however, such overall assessments will be based on performance data from several rounds of a scheme and will make provision for an occasional poor performance. Whether such disclosure is felt to be advantageous depends on the particular analytical sector, and would effectively be a decision of the steering committee, including representatives of the participants.

6.8 Integration of Proficiency Testing with Other Quality Assurance Measures

Proficiency testing is just one of a number of measures that may be used by analytical chemists in order to assure the quality of the data they are producing. It is important (Section 2.4) that laboratories adopt a comprehensive approach to quality assurance, employing the full range of measures available, rather than relying on proficiency testing alone.

The primary criterion for data quality is fitness for purpose – are the data sufficiently accurate to allow a user to make a correct decision based on them? Such a criterion can be expressed in terms of the uncertainty in an analytical result that is acceptable if that result is to be fit for purpose. The measures that the analytical chemist undertakes to ensure that the required uncertainty can be achieved come under the heading of quality assurance. These include good laboratory practices, use of validated and properly documented analytical methods, the use where appropriate of certified reference materials, proficiency testing and internal quality control. Figure 6.1 shows how these concepts and activities are connected.

Good laboratory practice (GLP) (in the general sense) comprises the basic activities that underlie the production of reliable data. It is concerned with appropriate levels in such matters as staff training and management, adequacy of the laboratory environment, the maintenance, testing and calibration of equipment, safety, the storage, integrity and identity of samples, record keeping, *etc.* Fulfilling these requirements is increasingly being examined by accreditation agencies.[38,39] Shortcomings in any of these features would tend to undermine vigorous attempts elsewhere to achieve the required quality of data. However, even in laboratories

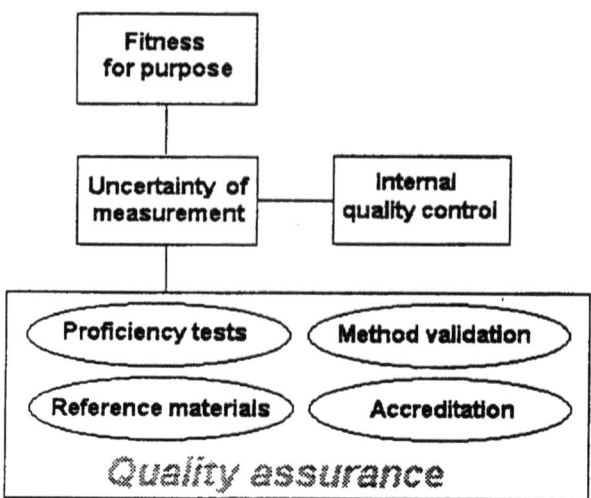

Figure 6.1 *Schematic diagram showing the relationship between the main concepts and practices relating to quality of data in analytical measurement. Internal quality control is normally regarded as part of quality assurance but plays a separate role, being retrospective*

where these conditions prevail, it is still possible for data of insufficient quality to be produced. Indeed there is no hard evidence at present to show whether accredited laboratories as a group perform better than others.

Validated methods of analysis provide the most obvious means of avoiding poor quality data. The validation process provides the evidence that the method is capable of performing to the required standard. The most rigorous (and expensive) means of method validation is the collaborative trial in which many laboratories participate.[40,41] Any difficulties in the interpretation of the method protocol and execution of the method will be revealed by the magnitude of the discrepancies between the results of the laboratories. The trial is carried out with a number of test materials representing the range of analyte concentration likely to be encountered in practice with the particular type of material. The method is characterised in terms of the repeatability precision, the reproducibility precision and (if certified reference materials are analysed) an estimate of the method bias with its uncertainty. Although there are thousands of methods that have been characterised by collaborative trial, this number still forms a small minority of the methods in use, for which a lesser degree of validation (usually within one laboratory) has been attempted. However, even when properly validated methods are being used, there is no guarantee that the documented performance characteristics that a method is capable of achieving will actually be achieved. This depends on circumstances in individual laboratories and on specific occasions. For example, methods might be used outside the scope of their validation, *i.e.* with inappropriate test materials. Internal quality control, the use of certified reference materials and proficiency testing are means by which the laboratory checks that it is using the methods correctly.

Internal quality control (IQC) is the means by which laboratories can ensure that with a high probability the data produced from day-to-day are of the required quality. The main recourse of IQC is the analysis of typical control materials of known analyte concentration among the test materials that make up the analytical run. The values obtained can be plotted on control charts[43] that reveal when the analytical system is 'out-of-control', and therefore producing data of unknown accuracy. In such circumstances the data should be rejected, remedial action taken on the aspects of the system causing the out-of-control condition, and only then the analytical run repeated. The IQC is an essential part of the routine running of an analytical laboratory.

Probably the main remedial effect of PT in improving data quality to the required level is achieved through the mechanism of encouraging laboratories to review and improve the effectiveness of their IQC procedures. It has been shown[14] that in one sector of the analytical community that laboratories with more comprehensive IQC programmes (independently assessed) perform significantly better in proficiency testing schemes. At the present time the efficacy of the IQC systems employed by laboratories often falls short of requirements, perhaps because until recently[13] there has been no clear guidance as to how such systems should be operated. Problems with IQC are often caused by the lack of suitable certified reference materials in certain measurement sectors. Such deficiencies in a laboratory's IQC programme are very efficiently revealed by proficiency testing.

Certified reference materials (CRMs) enable analysts to establish the traceability of their results to recognised measurement standards.[43] The accuracy of data and the comparability of data between different laboratories are thus enhanced, since laboratories are working to valid and common measurement standards. CRMs achieve this in two ways: (i) by providing reliable standard materials for calibration[44] and (ii) by providing matrix materials containing certified concentrations of the analytes sought in routine test materials.[45]

The use of a comprehensive range of quality assurance measures will ensure that all aspects of data quality are addressed. The particular benefits of proficiency testing (Section 6.1) will be realised only if it forms an integrated part of such a system. Without the support of a well-developed quality system, the benefits of participation in a PT scheme are much reduced.

CHAPTER 7

Proficiency Testing for End-users of Analytical Data

7.1 The Function of Proficiency Testing

Proficiency testing is a way of assessing the performance of analytical laboratories and, as a corollary, the quality of analytical data, by a series of regular inter-laboratory exercises. In a typical proficiency testing scheme test materials are distributed to test laboratories for analysis. The results are reported to the scheme organiser who then assesses the accuracy of each result by comparing the reported result with the assigned value. The laboratories are usually allocated a formal performance score for each test concerned. In certain types of scheme, where the test involved is non-destructive, a single specimen of the test material may be circulated to many participating laboratories for consecutive measurement. However, examples of such schemes are rare in the field of analytical chemistry. Normally a sample from a bulk material is distributed to each of a number of participants, and all of the samples are assumed to be essentially identical in composition.

The key feature of a proficiency test is therefore the regular and independent assessment of analytical data quality, using well-characterised test materials of proven homogeneity. Tests take place with a frequency between once a fortnight and once a year, with a typical frequency of once every 1–3 months. Laboratories involved are not informed of the assigned value until the proficiency exercise is completed.

PT schemes vary in detail of organisation, but all schemes share the common feature of the comparison of the results obtained by one laboratory to those obtained by one or more other laboratories. One, or a few, laboratories/organisations will have a controlling function. This may include characterisation of the test material to establish the assigned value for the material, against which the results of other laboratories will be judged. Alternatively, the assigned value may be based on the consensus of results from participating laboratories. Most proficiency tests also compare the accuracy of the results with a predetermined standard of performance, ideally based on fitness for purpose criteria (Section 7.2).

Although PT schemes could be organised for a single laboratory, in practice schemes usually have at least 10 participants, typically 50–100 and perhaps as many as 2000. PT schemes are usually organised on a self-financing basis and participants pay an annual participation fee.

Proficiency testing schemes usually operate on a declared or disclosed basis; that is the laboratories concerned know, while the analysis is being conducted, that they are taking part in a proficiency assessment. (Of course, participants do not know the assigned values of the analytes.) In special circumstances proficiency tests are organised on a non-disclosed basis and in this instance the PT materials are sent to the participating laboratories disguised as materials of a routine nature. Such covert trials are much more difficult and expensive to organise and operate than disclosed trials, but they do offer the advantage of providing a completely realistic assessment of routine laboratory performance.

Care must be taken not to confuse proficiency testing with the two other main types of interlaboratory exercise, namely *collaborative trials* (otherwise known as *method performance studies*) organised for the validation of analytical methods and *certification studies* organised for the characterisation of reference materials. It is important to distinguish these three activities because the objectives and operational procedures of each are quite different (Section 4.3). Because of these fundamental differences, it is not possible to combine the operation of a PT scheme with a formal method validation or reference material certification exercise. However, despite these differences it is sometimes possible to obtain subsidiary information on methods and materials from proficiency tests.

7.2 Fitness for Purpose

Analytical laboratories should produce data that are fit for purpose. Consequently, before commissioning a laboratory to carry out analytical work, end-users ideally should attempt to ascertain the accuracy they require from that analysis. Although it may be tempting to stipulate that an analytical result should be as accurate as possible, this is rarely a practical or cost-effective option. The more accurate a measurement, the more time and resource it will employ (for example, by involving a large number of replicate analyses or the use of specialist and expensive methodology). As laboratories clearly have to pass on the costs of increased accuracy, end-users can save on analysis costs by giving careful and realistic consideration to the accuracy they actually need in the data they are seeking.

For example, the end-user may need to enforce a maximum limit for contamination of 2 mg kg^{-1}. A typical measurement error of ±30% would not be acceptable where most sample results are expected to fall only slightly below this limit, but could be accepted for a screening procedure where most of the samples will be well below the contamination limit. Also, the consequences of making a wrong decision due to the errors of individual measurements must be considered. In some cases the consequences may be modest and the degree of confidence needed for each measurement can be correspondingly low. For example, in a large geochemical survey, it may be important to gather many results quickly and at low cost. The situation will be totally different for, say, clinical samples taken from individual patients. Here the consequences of a wrong decision could be life-threatening.

Little work has been done so far on finding exactly the maximum uncertainty that is acceptable in analytical data for a specific purpose. However, it has recently been demonstrated that a procedure of minimising a cost function could fulfil this need.[17]

7.3 Quality Procedures Necessary for Reliable Data

There are a number of key principles which laboratories can apply to help them to improve, maintain and demonstrate the validity of their data, namely the use of properly validated methods, the use of internal quality control procedures, the use of certified reference materials, accreditation and participation in proficiency testing schemes. When assessing which contract analytical laboratory to use, end-users may find it useful to establish which of these measures the laboratory has in place to ensure the quality of its data.

Each of the above measures makes its own particular contribution to analytical quality assurance and it is therefore important that they are used in conjunction with one another in order that shortcomings of one measure are offset by the use of other measures.

For example, while the use of internal quality control procedures ensures that a laboratory's results are consistent on a day-to-day basis, it does not always demonstrate that true results are being obtained. The IQC system might be flawed, with the result that the laboratory produces poor data consistently. The use of certified reference materials enables the trueness of data to be controlled, but suitable materials are not always available and even when they are, they may be expensive. Furthermore, as a laboratory knows the certified concentration in advance, CRMs do not provide a blind assessment and demonstration of data quality. Proficiency testing fulfils this role in that laboratories always analyse PT test material without prior knowledge of the assigned value of the test material. Also, as PT materials are distributed on a regular basis to participating laboratories, three to twelve times a year typically, a regular assessment of data quality is obtained.

7.4 How Proficiency Testing Can Help in the Selection of Reliable Laboratories

When an end-user is looking for a suitable contract laboratory to undertake analytical work, it should initially be established whether the laboratory participates in an appropriate proficiency testing scheme. Participation in itself demonstrates that the laboratory has a commitment to quality. It is important that end-users of analytical data are aware of the proficiency testing schemes that are in existence in the particular analytical sector in which they are involved. This will help them to assess whether a contract laboratory under consideration is using the available means of improving, maintaining and demonstrating quality. Consequently, a register of current UK proficiency testing schemes has been compiled[10] which gives information on the scope of each scheme (Appendix 1). Where a laboratory does not participate in an appropriate PT scheme, the reasons for non-participation should be sought before rejecting further consideration of that laboratory. Such laboratories are not necessarily indifferent to quality issues, but may simply find it too expensive to participate in all PT schemes relevant to their routine analytical work. Enquiries should be made as to what alternative quality measures are adopted by the laboratory to ensure an appropriate accuracy for their data, for example the use of certified reference materials or the use of spiked samples to test analyte recovery.

Where a laboratory does participate in a PT scheme the next step is to review its performance over several rounds. Certain schemes, on the basis of performance over several rounds, classify performance as, for example, 'acceptable' or 'unacceptable'.

Where such a classification is not produced by a scheme, performance in individual rounds may be evaluated. For those schemes using the z-score (see Section 3.2), an individual z-score for a particular test of between +2 and −2 is usually considered to indicate satisfactory performance. However, laboratories should not be judged on the basis of single z-scores. Criteria for classification of performance over several rounds are complex and are discussed at length in Sections 3.5, 3.6 and 5.18. Where a poor z-score has been obtained the laboratory should be able to provide evidence of appropriate and effective action being taken. Such evidence is indicative of a laboratory's commitment to quality.

Although schemes that are organised according to the guidelines in this book will utilise the z-score, other schemes may use a different scoring procedure. In such cases end-users will need to familiarise themselves with the particular scoring procedure adopted by obtaining a copy of the protocol produced by the scheme organiser.

7.5 Interpretation of Laboratory Performance in a Proficiency Testing Scheme

It is important to recognise that proficiency testing, like all quality measures, has inevitable limitations which need to be taken into account when reviewing and interpreting a laboratory's performance in a scheme. Particular consideration must be given to the conclusions that may reasonably be drawn from PT performance scores regarding the likely quality of a laboratory's daily output of data on *routine* test samples.

A number of factors need to be considered:

- Proficiency testing is retrospective in nature. It may be several weeks between the distribution of proficiency test materials through to the issue of performance assessments for a given round of a scheme. PT therefore reflects a laboratory's performance at a given point in the past.
- It is not always possible for scheme organisers to ensure that PT test materials are exact replicas of a laboratory's routine test materials. There may be differences in such features as the composition of the matrix, the concentration of the analytes and the form or speciation of the analytes.

Where such differences are significant, good performance in a PT scheme may not be indicative of good routine performance.

- The performance data provided by proficiency testing may be fairly specific for a particular test. For example, good performance in the analysis of zinc in foodstuffs does not necessarily indicate good performance in the analysis of calcium in foodstuffs, because interference effects from concomitant elements may vary in severity between analytes (see Figure 2.2). Caution must therefore be exercised when drawing conclusions regarding general laboratory performance on the basis of a few specific tests in a PT scheme.
- Participants in PT schemes are usually aware that they are being tested and therefore may be tempted to apply extra effort or more than usually stringent quality control measures to proficiency test materials in order to ensure that a good performance score is obtained.

7.6 Resource Implications of Proficiency Testing to End-users

A laboratory inevitably incurs costs when it participates in a proficiency testing scheme. Not only does a laboratory usually pay a fee to the scheme organiser, but there are costs associated with the staff time and consumables required to analyse and report the test samples. Obviously, these costs will be ultimately borne by the end-user in the form of increased charges for analytical services. It is difficult to estimate the contribution proficiency testing makes in routine analytical costs as it depends on the scale of the PT scheme and, more importantly, on the routine sample throughput of the laboratory. However, some indication has been obtained from the VAM Survey of PT schemes. For a laboratory with a high throughput (say 10 000 samples per year) proficiency testing perhaps adds 5% to the costs of analysis. However, for those laboratories undertaking fewer routine analyses, PT may contribute significantly to costs, perhaps approaching 25% or more. In these latter circumstances, stipulating that a laboratory participates in a scheme will inevitably incur a significant cost premium for the end-user. However, such premiums should be properly considered in the context of the possible financial consequences of using routine test data that are not fit for purpose. End-users should be aware of such consequences in their particular sector; in certain well-publicised instances liabilities exceeding millions of US dollars have been incurred as a result of inaccurate analysis. It is apparent, therefore, that the premium attached to quality represents a worthwhile investment for end-users of analytical data.

7.7 Accreditation Agencies

The accreditation of a laboratory's quality system by third party organisations is based upon such standards as ISO Guide 25[19] and European standard EN 45001.[46] These documents recognise the importance of proficiency testing activities as a means of demonstrating the quality of test results, particularly where the traceability of data to national measurement standards is difficult to establish.

Consequently, accreditation bodies such as UKAS[38] and CPA[39] strongly encourage candidate laboratories to participate in an appropriate PT scheme. In some instances participation is a mandatory requirement for accreditation. Where candidate laboratories were not participating in an available relevant scheme they would be expected to use a suitable alternative form of external quality control to demonstrate the accuracy of their data.

The occasional poor performance in a PT scheme should not, of itself, automatically lead to loss of accreditation, although an assessor would expect to see appropriate follow-up and corrective action taken. However, persistently poor performance over a number of rounds could well lead to loss of accredited status.

The precise way in which accreditation agencies make use of laboratory performance assessments from PT schemes depends on the agency and the scheme concerned. ISO Guide 43 on the Development and Operation of Laboratory Proficiency Testing,[9] originally issued in 1984, provided a little guidance on the role of proficiency testing in laboratory accreditation. However, this document is currently (June 1996) undergoing extensive revision, one aim of which is to provide much more comprehensive guidance on how proficiency testing should be used to assist the accreditation of test laboratories. Among the draft proposals are the following:

- PT schemes used by accreditation bodies should comply with the organisational guidelines laid down in ISO Guide 43;
- PT schemes should have appropriate criteria for judging successful performance in the proficiency test;
- the way the results of the proficiency test are used in accreditation decisions should be documented;
- accreditation should not be based on PT performance alone;
- accreditation bodies should have a documented procedure for acting on unsatisfactory results;
- accreditation bodies should consider the costs involved in proficiency testing.

For further information on the accreditation aspects of proficiency testing the reader should consult the revised version of ISO Guide 43,[47] due for publication in 1997.

7.8 Role of End-users in Scheme Organisation

Because of the common interest of end-users and scheme organisers in the assessment of analytical quality, there is ample scope for end-users to contribute to the organisation and design of PT schemes. Such contributions would help to ensure that scheme design takes account of the needs of end-users, which may differ from those of participating laboratories and scheme organisers. For example, end-users might wish one output of a PT scheme to be a publication listing all laboratories achieving satisfactory performance in the scheme. Although schemes organised according to the principles stated in this book could produce such a listing, few schemes currently operating actually do so and, indeed, many participants would feel that such a policy would deviate from the self-help ethos of proficiency testing. End-user representation on a scheme's steering committee would enable the end-user viewpoint on these and other topics to be discussed. Details of the constitution of the scheme's steering committee would be obtainable from the contact point for the scheme organiser.

CHAPTER 8
Proficiency Testing – The Future

While the VAM survey of proficiency testing was being undertaken, it soon became apparent that proficiency testing was an area undergoing rapid developments. For example, schemes were growing in size and many were operating increasingly on an international basis. The range of analytical tests covered by schemes was growing as was the regulatory and accreditation influences on proficiency testing. The costs (and benefits) of proficiency testing were being scrutinised in more detail by organisers and participants alike and new developments in computer software were being introduced by several schemes to cope with the increasing amount of data, both technical and administrative, that was being generated. For reasons such as these, it is appropriate to conclude the present book with a brief 'forward look' at some of the issues likely to be of importance to proficiency testing in the future.

The comparability of analytical data on an international basis is of major importance and PT activities in the future are likely to reflect this. Currently consideration is being given to the establishment, through the Standards Measurement and Testing Programme of the European Union, of a pan-European proficiency testing network covering measurements of importance to the enforcement of EU directives. Certain of these directives, for example those relating to the control of foodstuffs, require laboratories undertaking analyses to participate in an appropriate PT scheme. Such requirements are also likely to become a feature of other test sectors. The identification of an appropriate scheme by analytical laboratories requires access to an up-to-date listing of existing schemes. Listings of this type, in the form of a computerised database covering European schemes, are currently being developed at the Laboratory of the Government Chemist as part of the VAM Initiative.

As proficiency testing schemes assess the quality of analytical data, it is vital that the schemes themselves are operated effectively and meet specific quality criteria. To this end proficiency schemes are increasingly seeking certification of their operational procedures to recognised quality standards, such as the ISO 9000 series. To assist this process work is underway at the LGC to identify and document appropriate quality systems and audit protocols for PT schemes.

The effectiveness of formal quality assurance measures in improving the quality of analytical data is a topic of concern, particularly in respect of the growing QA requirements placed on laboratories and the cost burden that entails. In this context, a laboratory's performance in a PT scheme provides an indication of its competence and consequently correlations between PT performance and other aspects of a laboratory's quality system (*e.g.* do accredited laboratories perform better than non-accredited laboratories?) are increasingly being sought. A related issue concerns the impact of quality assurance on a laboratory's routine output of data. A pertinent question that proficiency testing schemes may well face in the future could be 'do laboratories that perform well in a PT scheme (where they know they are being tested) necessarily perform equally well in routine analysis?' The question is by no means academic as one of the findings of the VAM survey on proficiency testing was that end-users of analytical data do sometimes experience quality problems with routine sample data despite using a laboratory that has performed satisfactorily in a PT scheme.

Uncertainty of chemical measurements is a topic that has come to the fore in recent years. Procedures for handling uncertainty estimates attaching to analytical results have not been extensively developed by PT schemes, but such procedures are likely to appear and ultimately participants in schemes will be expected to quote an uncertainty for all the results they report to the scheme organiser.

This book has dealt entirely with proficiency testing as it is applied to *quantitative* analysis. In certain areas, however, for example forensic investigations and medical diagnoses, a *qualitative* test may be performed, the result of which may be used to formulate a professional opinion. Currently there is no generally recognised and agreed way of assessing proficiency in such work, despite its great importance. Efforts to establish appropriate proficiency testing procedures for qualitative tests can be expected in the coming years.

Any measurement, whether qualitative or quantitative, must be made in the context of being 'fit for purpose', that is the measurement must be of an accuracy that enables an end-user of the measurement to make a sound technical, commercial, regulatory or other decision. Although fitness for purpose is widely spoken about in general terms, no commonly accepted procedures exist for establishing in specific terms what actually constitutes fitness for purpose. Such procedures are clearly required and it is to be hoped that these will be devised and incorporated into the operational protocols of PT schemes.

Appropriate quality of sampling is a topic that is important for analytical chemistry, and has to be considered alongside fitness for purpose in analysis. The quality of the result cannot possibly be better than the quality of the sample. As sampling quality is in many aspects formally analogous with analytical quality, it is tempting to consider the idea of sampling proficiency tests.[48] A pilot study (in the contaminated land sector of activity) has indeed shown that the idea is practicable,[49] but it remains to be seen whether the idea will 'catch on'.

Finally, readers will be interested to note that an international collaborative activity to revise ISO Guide 43, which deals with proficiency testing, is due for completion in 1997. The new Guide greatly extends the scope of the original version published in 1984, providing detailed guidance on such matters as the use of proficiency testing by accreditation bodies and statistical procedures for handling proficiency test data, as well as the organisational procedures recommended for use in PT schemes.

Glossary of Terms and Acronyms

This glossary is intended to provide a brief explanation of those technical terms and acronyms that are used in this book, but which may not be familiar to the reader. The explanations given are not neccessarily formal or official definitions. Cross-references are identified by italics.

Accreditation (of a Laboratory). Laboratory accreditation is formal recognition by an authorised or recognised body that a testing laboratory is competent to carry out specific tests or specific types of tests.

Accuracy of Measurement. The closeness of the agreement between the result of a measurement and the *true value* of the *measurand*.[34]

Analyte. The component of a test material which is ultimately determined.

AOAC International. A USA-based organisation previously called the Association of Official Analytical Chemists.

Assigned Value. The assigned value for an *analyte* in a test material distributed in a proficiency testing scheme is the value adopted by the scheme organiser as the best available estimate for the concentration of the analyte in question. For a proficiency testing scheme to be of value the assigned value must be a good approximation to the *true value*.

AQUACHECK. A proficiency testing scheme operated by the Water Research Centre, Medmenham, UK and covering the analysis of waters, soils and sludges.

BAPS. The Brewing Analytes Proficiency Scheme, operated by the Laboratory of the Government Chemist and dealing with the analysis of beers.

Bias. Characterises the *systematic error* in a given analytical procedure and is the (positive or negative) deviation of the mean analytical result from the *true value*.[50]

BS. British Standard.

Certified Reference Material (CRM). A *reference material*, accompanied by a certificate, one or more of whose property values are certified by a procedure which establishes its *traceability* to an accurate realisation of the unit in which the property values are expressed, and for which each certified value is accompanied by an *uncertainty* at a stated level of confidence.[30]

CHEMAC. Formerly the Chemical Measurement Advisory Committee, the work of this body has now been subsumed into the activities of *Eurachem-UK*.

CITAC. Cooperation on International Traceability in Analytical Chemistry, an international group that aims to foster worldwide collaboration on the mechanisms needed to ensure the validity and comparability of analytical data on a global basis. CITAC may be contacted through the CITAC Secretariat, Laboratory of the Government Chemist, Queens Road, Teddington, Middlesex, TW11 0LY, UK.

Collaborative Trial. An analytical study involving a number of laboratories analysing representative portions of the same test material and using the same method, for the purpose of validating the performance of the method. This approach is often used to establish the *trueness* and/or the *precision* of an analytical method. The term is often used interchangeably with method performance study.

Consensus Value (Participant). In this book, the participant consensus value is the mean of participants' results on a test material distributed in a proficiency testing scheme, after *outliers* have been handled either by elimination or by the use of *robust statistics*. The participant consensus value thus calculated may be adopted as the *assigned value* for the material. Where the results deviate from a *normal distribution*, expert judgement may be needed to establish a consensus.

CONTEST. A proficiency testing scheme operated by the Laboratory of the Government Chemist and covering the analysis of contaminated land.

Control Material. A control material is used for the *quality control* of routine analyses, by inclusion in a run of test samples undergoing analysis.[13]

Coverage Factor. Numerical factor used as a multiplier of the combined *standard uncertainty* in order to obtain an *expanded uncertainty*.[51]

CPA. Clinical Pathology Accreditation (UK) Ltd., a body undertaking the *accreditation* of clinical test laboratories.

EN. Euro Norm or European Standard

EQAS. External quality assessment scheme, sometimes used interchangeably with *proficiency testing*.

Error. The error of an analytical result is the difference between the result and the *true value.*

EURACHEM. A pan-European organisation that aims to promote the validity and comparability of analytical data on a Europe-wide basis.

Eurachem-UK. The UK's national *EURACHEM* organisation. It aims to help laboratories improve the quality of analytical data and provides UK input to EURACHEM. It is assisted by specialist working groups, including one dealing with proficiency testing. The Eurachem-UK Secretariat may be contacted at PO Box 46, Teddington, Middlesex, TW11 0NH, UK.

Expanded Uncertainty. Quantity defining an interval about the result of a measurement that may be expected to encompass a large fraction of the distribution of values that could reasonably be attributed to the *measurand.*[51]

Extreme Results. *Outliers* and other values which are grossly inconsistent with other members of a set of results.[47]

FAPAS. The Food Analysis Performance Assessment Scheme, covering a wide range of food analyses and operated by the UK Ministry of Agriculture, Fisheries and Food.

Fitness for Purpose. Analytical data must be of an *accuracy* that is sufficient to enable an end-user of the data to make sound decisions. Such data are said to be fit for the intended purpose. In order to establish whether a result is fit for purpose, the *uncertainty* attaching to the result must be documented. It should be noted that data do not necessarily have to be of the highest accuracy possible in order to fulfil the fitness for purpose criterion.

GeoPT. A newly established PT scheme covering the analysis of silicate rocks for about 50 elements.

IUPAC. The International Union of Pure and Applied Chemistry.

ISO. The International Organisation for Standardisation.

LEAP. Laboratory Environmental Analysis Proficiency, a UK-based PT scheme for environmental analysis operated by Yorkshire Environmental.

MAFF. Ministry of Agriculture, Fisheries and Food (UK).

Measurand. Particular quantity subject to measurement.[34]

Median (of a data set). If the data are arranged in order, the median is the central number of the series, *i.e.* there are equal numbers of observations smaller and greater than the median.

Median of the Absolute Deviations (MAD). When the absolute deviations (*i.e.* ignoring the sign of the deviations) of the individual members of a data set from the *median* of that set are arranged in order of magnitude, the median of the absolute deviations (MAD) may be obtained. By multiplying the MAD value by the factor 1.483 a *robust* standard deviation of the data set is obtained. See also *robust statistics*.

Mode (of a data set). The value which occurs most frequently, *e.g.* for the data set 1,2,2,3,3,3,3,4,4,5,6, the mode is 3.

NACS. National Agricultural Check Sample Scheme, a UK proficiency testing scheme for animal feedstuffs operated by Perstorp Analytical.

NAMAS. The National Measurement Accreditation Service, now known as the United Kingdom Accreditation Service (UKAS), a body that accredits test laboratories to quality standards such as EN 45001[46] (ISO Guide 25).

NEQAS. The UK National External Quality Assessment Scheme(s), a network of about 35 schemes operating from about 20 organising centres and covering a wide range of biomedical and related analyses. Operated in conjunction with the UK Department of Health.

Normal Distribution. One of the standard distributions of statistical theory. Errors of measurement often approximate to a normal distribution.

Outlier. A member of a set of results which is statistically inconsistent with the other members of that set.[47]

PHLS. The UK Public Health Laboratory Service.

Precision. The closeness of agreement between the results obtained by applying an analytical method to the same sample several times under prescribed (*repeatability*) conditions.

Proficiency Testing (PT). Study of laboratory performance by means of ongoing interlaboratory test comparisons.[47] [Proficiency testing schemes are also referred to as external quality assessment schemes (EQAS) or laboratory performance studies in some test sectors.]

ProTAS. Proficiency Testing for Alcoholic Strength, a PT scheme for wines and ciders operated by the Laboratory of the Government Chemist.

Quality Assurance (QA). The sum total of a laboratory's activities aimed at achieving the required standard of analysis. In addition to the QC procedures applied by laboratory staff, QA also includes such matters as staff training, administrative procedures, management structure, auditing, *etc.*[7]

Quality Control (QC). The set of procedures undertaken by laboratory staff for the continuous monitoring of the results of measurements in order to decide whether results are reliable enough to be released.[7] Distinction is sometimes made between internal quality control and external quality control. The former is concerned with monitoring the *repeatability* of results, for example by carrying out regular replicate analyses on a material known to be homogeneous, such as a *control material*. External quality control monitors the *accuracy* of results and involves, for example, the use of *certified reference materials* and participation in *proficiency testing schemes.*

Random Error (*Precision*). An error which varies in an unpredictable manner in absolute value and in sign when a large number of measurements are made on the same sample.

Reference Material. A material or substance one or more of whose property values are sufficiently homogeneous and well-established to be used for the calibration of an apparatus, the assessment of a measurement method or for assigning values to materials.[30]

Repeatability Conditions. A series of measurements (on identical portions of the same sample for the same analyte) are said to have been carried out under repeatability conditions when the same analyst applies the same measurement procedure, using the same equipment, in the same location, over the shortest practical timescale.

RICE. The Regular Interlaboratory Counting Exchanges, a UK-based proficiency testing scheme operated by the Institute of Occupational Medicine and dealing with asbestos fibre counting.

Robust Statistics. Statistical techniques used to minimise the influence that *extreme results* can have on estimates of the mean and *standard deviation* (or other statistics) of a set of results. These techniques assign less weight to extreme results, instead of eliminating them entirely from a dataset. The *median* is a robust estimate of the mean of a set of results. The robust standard deviation of the results may be calculated from the *MAD (median of the absolute deviations)* value.[16]

Spiked Samples. 'Spiking a sample' is a widely used term taken to mean the addition of a known quantity of *analyte* to a sample matrix which is close to or identical with the samples of interest. (Also referred to as fortified samples.)

Standard Deviation. A measure of the dispersion of data about the mean value. Mathematically it is equal to the square root of the *variance* of the results.

Standard Uncertainty. *Uncertainty* of a measurement expressed as a *standard deviation.*[51]

Systematic Error (Bias). An error which remains constant in absolute value and sign when a number of measurements are carried out on the same sample.

Target Standard Deviation. A numerical value, used in the calculation of a *z-score*, that has been designated by the organiser of a *proficiency testing scheme* as a goal for measurement quality.[7]

Traceability. The property of the result of a measurement or the value of a standard whereby it can be related to stated references, usually national or international standards, through an unbroken chain of comparisons, all having stated *uncertainties*.[34]

True Value. The actual concentration of an *analyte* in a sample matrix. It is the value that would be obtained by a perfect measurement. (True values are by nature indeterminate.)[34]

Trueness. The closeness of the agreement between the mean of a large number of results (obtained by applying an analytical method to the same sample under *repeatability conditions*) and the *true value* for the sample in question.[11]

VAM Initiative. Valid Analytical Measurement Initiative, a programme funded by the UK's Department of Trade and Industry, aimed at improving the quality of analytical data and facilitating the mutual acceptance of analytical data on both a national and an international basis.

Variance. A measure of the dispersion of individual analytical results about the mean of those results.

WASP. The Workplace Analysis Scheme for Proficiency, a UK proficiency testing scheme covering toxic substances in the workplace atmosphere and operated by the Health and Safety Executive.

u_a. The symbol used in this book to represent the *uncertainty* of the *assigned value* of a test material used in a PT scheme.

u_x. The symbol used in this book to represent the *uncertainty* of an analytical result.

Uncertainty (of measurement). A parameter associated with the result of a measurement that characterises the dispersion of the values that could reasonably be attributed to the *measurand*.[34]

UKAS. United Kingdom Accreditation Service (see also *NAMAS*).

x. The symbol used in this book to represent an analytical result.

x_a. The symbol used in this book to represent the *assigned value* of a test material used in a PT scheme.

z-Score. A performance score recommended for use in proficiency testing schemes to evaluate the accuracy of the analytical results submitted by participating laboratories:[7]

$$z\text{-score} = \frac{\{\text{Laboratory Result} - \text{Assigned Value}\}}{\text{Target Standard Deviation}}$$

Appendix 1: Eurachem-UK Register of Proficiency Testing Schemes

The UK proficiency testing schemes listed in the latest version of the register of schemes produced by the Eurachem-UK Proficiency Testing Working Group are shown below. Full details are held on a computerised database maintained by the Laboratory of the Government Chemist, from whom up-to-date information on contact names, addresses, *etc.* for the various organising bodies may be obtained. Further explanation of the acronyms used may be found in the glossary.

NAME OF SCHEME	SCOPE	ORGANISING BODY
Occupational Hygiene		
MMMF Counting Exchanges	Man-made mineral fibre counting	Institute of Occupational Medicine
RICE	Asbestos fibre counting	Institute of Occupational Medicine
WASP	Hazardous substances in the workplace atmosphere	Health and Safety Executive
Food and Agriculture		
BAPS	Analysis of beers	Laboratory of the Government Chemist
DAPS	Analysis of distilled spirits and fermented worts	Laboratory of the Government Chemist
FAPAS	Food Analysis Performance Assessment Scheme	Ministry of Agriculture, Fisheries and Food
NACS	Animal feedstuffs	Perstorp Analytical
PHLS EQAS	Microbiological examination of foods	PHLS Food Hygiene Laboratory
ProTAS	Alcoholic strength of wines and ciders	Laboratory of the Government Chemist
Quality in Microbiology Scheme	Microbiological examination of foods	Quality Management

NAME OF SCHEME	SCOPE	ORGANISING BODY
QAS	Microbiological examination of foods	Ministry of Agriculture, Fisheries and Food
—	Animal feedstuffs, soil compost, hydroponic solutions	Agricultural Development Advisory Service
—	Flour and wheat (functional properties)	Flour, Milling and Baking Research Association
—	Flour and wheat (NIR and Kjeldahl N)	Flour, Milling and Baking Research Association
—	Analysis of cane sugar	Sugar Association of London

Water and Environment

AQUACHECK	Waters, soil and sludge analysis	WRc, Medmenham
CONTEST	Toxic contaminants in soil	Laboratory of the Government Chemist
Environmental Radioactivity	Alpha, beta and gamma emitting radionuclides	National Physical Laboratory
LEAP	Water, industrial effluent, cryptosporidium, water microbiology, toxic contaminants in soil	Yorkshire Environmental
PHLS EQAS	Water microbiology, indicator organisms, cryptosporidium, legionella	PHLS

Biomedical

Murex Quality Assessment Programme	Clinical chemistry and biomedical analyses	Murex Diagnostics
UK Cyclosporin QA Scheme	Cyclosporin in blood	St George's Hospital

NAME OF SCHEME	SCOPE	ORGANISING BODY
UK NEQAS	Clinical chemistry, haematology, immunology, histopathology and microbiology investigations	UK NEQAS (Department of Health)
Veterinary Laboratory Quality Assessment Scheme	Biochemistry, bacteriology, virology and animal tissue analyses	MAFF Veterinary Investigation Service
WEQAS	Clinical chemistry	University Hospital of Wales

Industry

InterCentre Precision Monitoring Scheme	Petroleum products	British Petroleum
Quality Counts	Manual colony counting using a spiral plater	Don Whitley Scientific Ltd.

Forensic

Forensic QA	Forensic investigations	Home Office Forensic Science Laboratory

Appendix 3: Examples of Performance Reports Issued by PT Scheme Organisers

The following illustrations show examples of typical performance reports issued by five PT schemes to participant laboratories.

The schemes concerned are: UK NEQAS for clinical chemistry, the Workplace Analysis Scheme for Proficiency (WASP), the Regular Interlaboratory Counting Exchanges (RICE), the Food Analysis Performance Assessment Scheme (FAPAS) and the Brewing Analytes Proficiency Scheme (BAPS).

The illustrations are reproduced by kind permission of the schemes concerned.

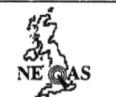

UK NEQAS for Clinical Chemistry	Laboratory :
Distribution : 523 Date : 9-Oct-95	Page 2 of 14
Specimen : 523	

Sodium (mmol/L)

	n	Mean	SD	CV(%)
All methods	475	154.89	2.29	1.5
Flame/integral dilutor	24	155.33	1.84	1.2
Indirect ISE	321	154.90	2.11	1.4
Technicon	40	154.67	2.26	1.5
Beckman	88	155.33	1.90	1.2
Other	83	154.58	2.08	1.3
Bayer DAX/Axon	26	153.53	1.74	1.1
Hitachi 704/717/737/747/911	67	155.34	2.12	1.4
ILab 900/1800	8	155.48	2.31	1.5
Kodak Ektachem system	67	154.39	1.79	1.2
EK700/750	53	154.25	1.83	1.2
Other method	61	155.28	3.68	2.4
Direct ISE - other manufact.	34	154.81	4.29	2.8

☐ All methods
■ Indirect ISE
■ ILab 900/1800

Your result	158
Target value (MLTM)	154.90
Your bias (%)	+2.0
Your BIS	+125
Your MRVIS	76
Your MRBIS	17
Your SDBIS	85

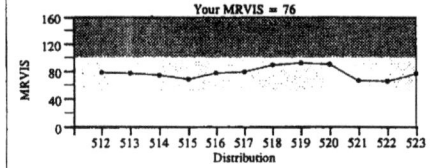

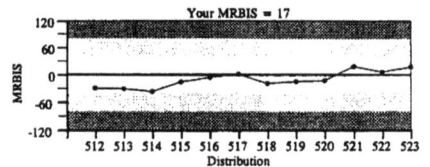

Potassium (mmol/L)

	n	Mean	SD	CV(%)
All methods	476	5.08	0.12	2.3
Flame/integral dilutor	24	5.07	0.12	2.3
Indirect ISE	322	5.08	0.10	2.0
Bayer SMAC/RA system	40	5.05	0.12	2.3
Beckman	88	5.12	0.09	1.7
Other	83	5.06	0.10	1.9
Bayer DAX/Axon	27	5.04	0.11	2.2
Hitachi 704/717/737/747/911	67	5.09	0.07	1.4
ILab 900/1800	8	5.10	0.11	2.1
Kodak Ektachem system	67	5.04	0.11	2.2
EK700/750	53	5.04	0.12	2.4
Other method	61	5.10	0.15	2.9
Direct ISE - other manufact.	34	5.08	0.14	2.8

☐ All methods
■ Indirect ISE
■ ILab 900/1800

Your result	5.3
Target value (MLTM)	5.08
Your bias (%)	+4.3
Your BIS	+147
Your MRVIS	48
Your MRBIS	42
Your SDBIS	58

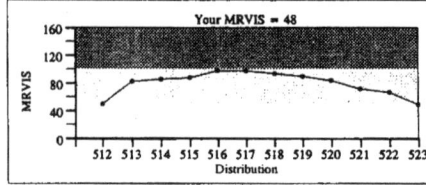

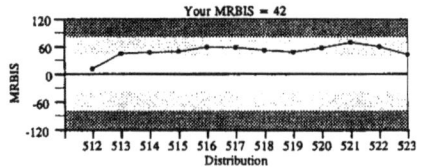

UK NEQAS, Wolfson EQA Laboratory,
PO Box 3909, BIRMINGHAM B15 2UE, UK
Phone (direct) 0121-414 7300;   FAX 0121-414 1179

WASP Results   Round: 27   Laboratory: 149

Toluene on Charcoal Tubes

Frequency Count of Results for this Analyte

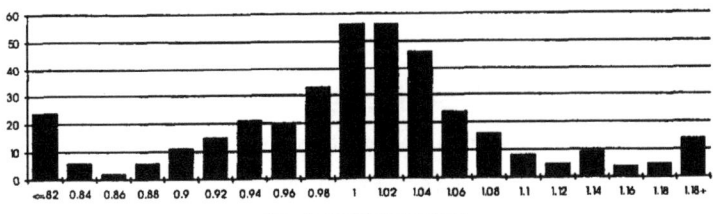

Results for this Laboratory

summary of your results and their ratio to the mean

sample number	your result	'true' result	standarised results		RPI	performance category
1	256.00	242.95	1.054		< 16	1
2	77.00	81.02	0.950		16 to 65	2
3	378.00	359.24	1.052		> 65	3
4	193.00	190.35	1.014			
		mean:	1.02		*RPI Reference Value:* 36	

Number of outliers:	0		
Performance Index (PI):	21	Mean PI: 37.10	Median PI: 28
RunningPerformanceIndex (RPI):	5	Mean RPI: 36.82	Median RPI: 24
Performance Category:	1	Your laboratory is ranked 13 out of 88 labs.	

additional statistical data (see the statistical protocol paragraphs 12 to 18)

analysis of variance: **Delta:** 0.017577   **SSW:** 0.007029

RPI plot for your laboratory

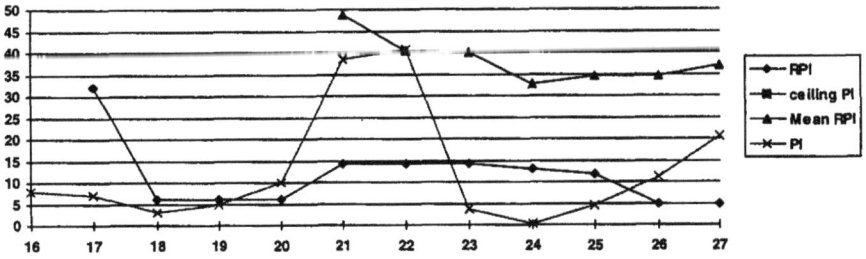

Doc.No. : CL1/A
IOM File: QA P.523
Ref : ***/31-34

C ommittee on **F** ibre **M** easurement

RICE Scheme

(Quality Control of Asbestos Fibre Counting)

CLASSIFICATION OF LABORATORY PERFORMANCE

Period   :   September 1994 – October 1995

Rounds   :   31 – 34

Laboratory   LABORATORY A
1 First Street
Sometown
ZZ1 1ZZ

Classification   :   SATISFACTORY

No. of rounds   :   4
completed in above period

Notes

1   The classification will be updated after each subsequent round ie at 3-4 monthly intervals.

2   Laboratories which have completed fewer than four rounds are classified as "AWAITING CLASSIFICATION".
On completion of four rounds laboratories will be classified as "SATISFACTORY" or "UNSATISFACTORY".

Institute of Occupational Medicine, Edinburgh

11 November 1995

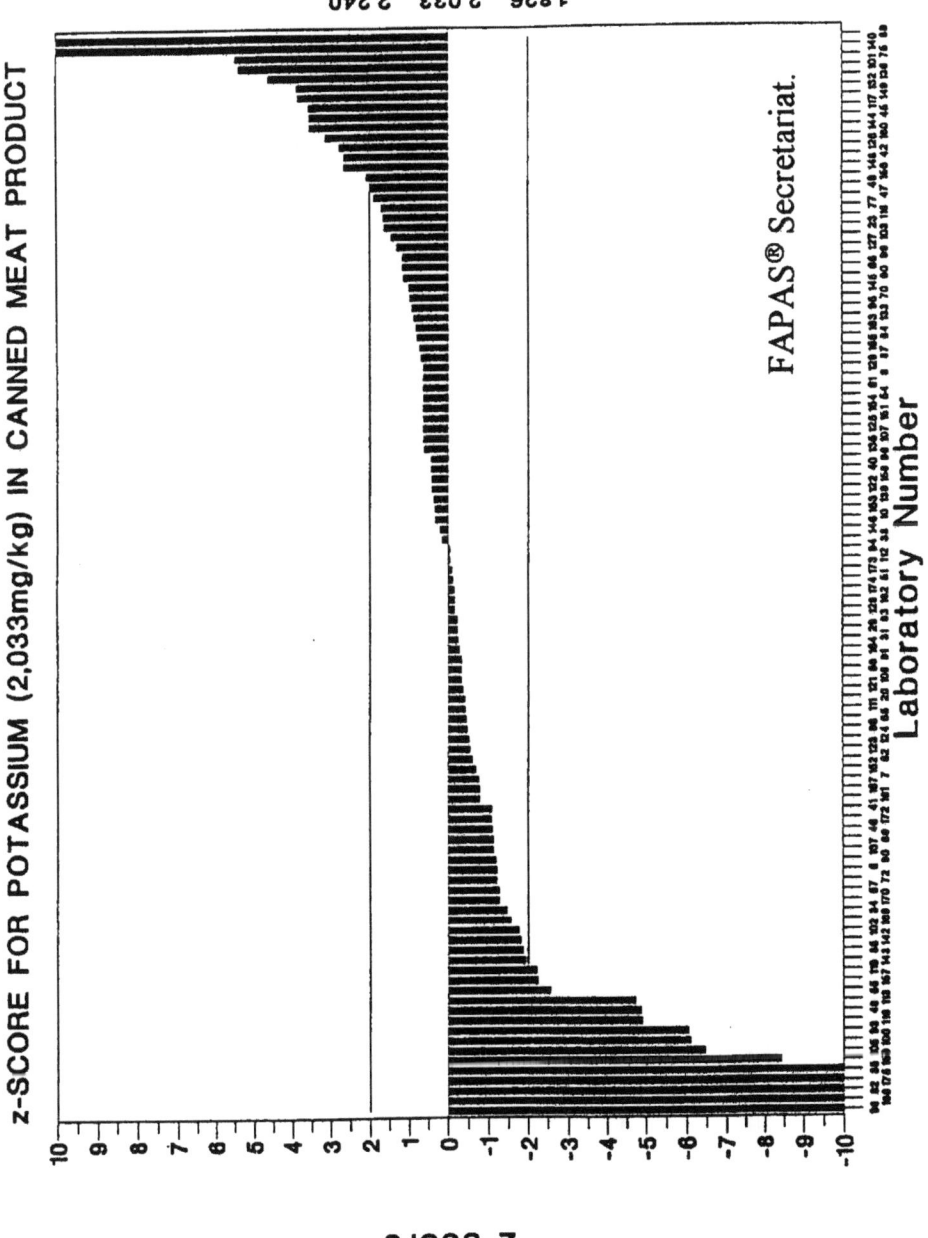

z-SCORE FOR POTASSIUM (2,033mg/kg) IN CANNED MEAT PRODUCT

FAPAS® Secretariat.

10/07/95

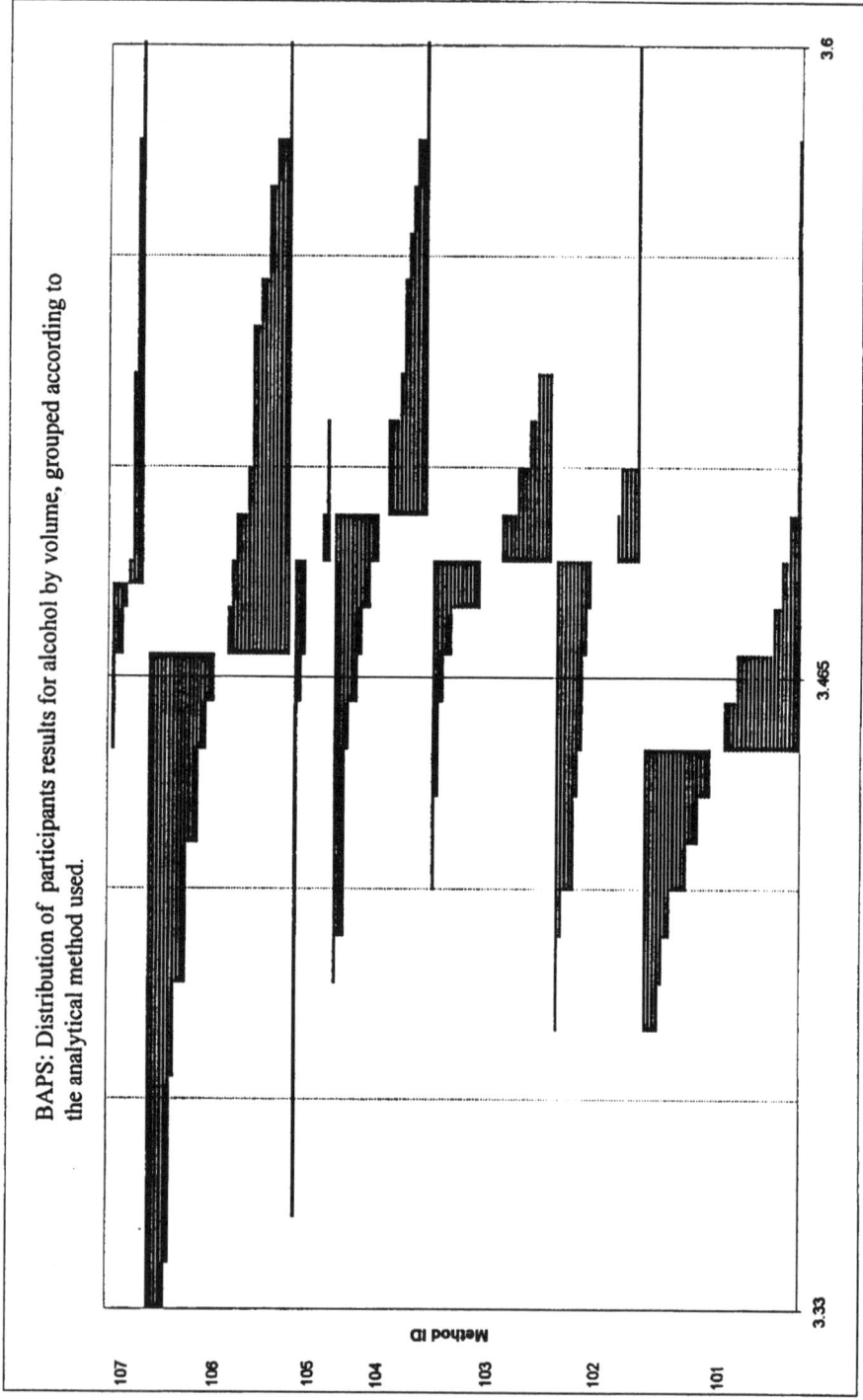

BAPS: Distribution of participants results for alcohol by volume, grouped according to the analytical method used.

References

1. W. Horwitz, L.R. Kamps and K.W. Boyer, *J. Assoc. Off. Anal. Chem.*, 1980, **63**, 1344.
2. N.W. Hanson, 'Official, Standardized and Recommended Methods of Analysis', Society for Analytical Chemistry, London, 1973.
3. W.B. Hamlin, *Clin. Chem.*, 1993, **39**, 1456.
4. D.G. Bullock, 'External Quality Assessment in Clinical Chemistry – An Examination of Requirements, Applications and Benefits', Ph.D. Thesis, 1987, University of Birmingham.
5. T.P. Whithead, D.M. Browning and A. Gregory, *J. Clin. Pathol.*, 1973, **26**, 435.
6. J.C. Sherlock, W.H. Evans, J. Hislop, J.J. Kay, R. Law, D.J. McWeeny, G.A. Smart, G. Topping and R. Wood, *Chem. Br.*, 1985, **21**, 1019.
7. M. Thompson and R. Wood, *J. AOAC Int.*, 1993, **76**, 926.
8. RSC Analytical Methods Committee, *Analyst*, 1992, **117**, 97.
9. ISO Guide 43: 1984, 'Development and Operation of Laboratory Proficiency Testing'.
10. Anon., *Anal. Proc. Anal. Commun.*, 1994, **31**, 285.
11. ISO 3534-1: 1993, 'Statistics – Vocabulary and Symbols – Part 1: Probability and General Statistical Terms'.
12. N.W. Tietz, D.O. Rodgerson and R.H. Laessig, *Clin. Chem.*, 1992, **38**, 473.
13. M. Thompson and R. Wood, *Pure Appl. Chem.*, 1995, **67**, 649.
14. M. Thompson and P.J. Lowthian, *Analyst*, 1993, **118**, 1495.
15. D. McCormick and A. Roach, 'Measurements, Statistics and Computation', Chapter 3, p. 114, in 'Analytical Chemistry by Open Learning', J. Wiley and Sons Ltd., 1987.
16. RSC Analytical Methods Committee, *Analyst*, 1989, **114**, 1693.
17. M. Thompson and T. Fearn, *Analyst*, 1996, **121**, 275.
18. ISO 9001: 1987, 'Quality Systems – Model for Quality Assurance in Design/Development, Production, Installation and Servicing'.
19. ISO Guide 25: 1990, 'General Requirements for the Competence of Calibration and Testing Laboratories'.
20. D.G. Bullock, N.J. Smith and T.P. Whitehead, *Clin. Chem.*, 1986, **32**, 1884.
21. Anon., *Pure Appl. Chem.*, 1994, **66**, 1903.
22. W. Horwitz, *Anal. Chem.*, 1982, **54**, 67A.
23. N.P. Crawford, P. Brown and A.J. Cowie, *Ann. Occup. Hyg.*, 1992, **36**, 59.
24. H.M. Jackson and N.G. West, *Ann. Occup. Hyg.*, 1992, **36**, 545.
25. B. Tylee, Health and Safety Executive, Sheffield, Personal Communication, 1992.
26. P.W. Britton, United States Environmental Protection Agency, Cincinnati, Personal Communication, 1992.
27. J.P. Weber, *Sci. Total Environ.*, 1988, **71**, 111.
28. M. Thompson, R.E. Lawn and R.F. Walker, unpublished work.
29. ISO Guide 35: 1989, 'Certification of Reference Materials – General and Statistical Principles'.
30. ISO Guide 30: 1992, 'Terms and Definitions Used in Connection with Reference Materials'.
31. V. Barnett and T. Lewis, 'Outliers in Statistical Data', Wiley, New York, 1984.

32. J. Seth, I. Hamming, R.R.A. Bacon and W.M. Hunter, *Clin. Chem. Acta*, 1988, **174**, 171.
33. P.J. Huber, 'Robust Statistics', Wiley, New York, 1981.
34. ISO, 'International Vocabulary of Basic and General Terms in Metrology', 2nd Edn., 1993 (ISBN 92-67-01705-1).
35. EURACHEM, 'Quantifying Uncertainty in Analytical Measurement', 1995 (ISBN 0-948926-08-2).
36. R.W. Jenny and K.Y. Jackson, *Clin. Chem.*, 1992, **38**, 496.
37. Anon., *The Referee*, 1990, **14**, 6.
38. UKAS Executive, 'Accreditation for Chemical Laboratories', edition 1, NIS 45, 1990.
39. Anon., *J. Clin. Pathol.*, 1991, **44**, 798.
40. W. Horwitz, *Pure Appl. Chem*, 1995, **67**, 331.
41. ISO 5725 (Parts 2 and 4): 1994, 'Accuracy (Trueness and Precision) of Measurement Methods and Results'.
42. E. Mullins, *Analyst*, 1994, **119**, 369.
43. J.K. Taylor, 'Quality Assurance of Chemical Measurements', Chapter 18, Lewis Publishers Inc., 1987.
44. Ref. 43, Chapter 10.
45. ISO Guide 33: 1989, 'Uses of Certified Reference Materials'.
46. EN 45001: 1989, 'General Criteria for the Operation of Testing Laboratories'.
47. ISO Guide 43: 1997, 'Proficiency Testing by Interlaboratory Comparisons'.
48. M. Thompson and M.H. Ramsey, *Analyst*, 1995, **120**, 261.
49. A. Argyraki, M.H. Ramsey and M. Thompson, *Analyst*, 1995, **120**, 2799.
50. A.D. McNaught and A. Wilkinson, 'Compendium of Chemical Terminology IUPAC Recommendations', 2nd Edn., Blackwell Science, 1997.
51. ISO, 'Guide to the Expression of Uncertainty in Measurement', 1993. (ISBN 92-67-10188-9).

Subject Index